Home Attitude

EVERYTHING YOU NEED TO KNOW TO MAKE YOUR HOME SMART

D0111830

John R. Patrick

Attitude
LLC

ISBN: 1542710197
ISBN 13: 9781542710190

Praise for *Home Attitude*

"Home Attitude is your practical guide to take advantage of the many benefits of home automation. You will be amazed at what the future will bring, but even more about what's possible today. Futurist John Patrick points the way for all of us."

Skip Prichard, President & CEO, OCLC, Inc. and author of *The Book of Mistakes*

"John Patrick's latest book in his growing Attitude Series provides an excellent overview of how a hands-on homeowner can build an automated home which is energy efficient, secure, and just plain fun. Patrick clearly explains the many benefits of an automated home, its key building blocks, and how to get started. Whether you're interested in simply reducing your utility bills, or building the home of the future, *Home Attitude* will help make it possible today."

Ronald H. Gruner, Founder, Alliant Computer and Shareholder.com

"John Patrick is a renowned visionary and evangelist for what is possible when the Internet, technology, and people intersect. *Home Attitude* brings John's insightful principles and thought leadership intimately into our homes. This is a must read for anyone feeling their home is falling behind the technology trends of our time."

Dan Ohlson, Founder, Realtek Holding Investments

"John is one of the very few people who have been a driving force behind the PC and Internet revolution. Remember Mark Zuckerberg's smart home video with JARVIS? It is no longer rocket science to automate your home. Anyone with the right attitude and dedication can build a smart home for a very reasonable price. Read John's book to gain this attitude and let your home make your life easier."

Bilal Athar, CEO, Wifigen LLC

"Automating your home is much more than a hobby," notes Dr. John. "It is an attitude." "It is this state of mind, coupled with his ability to sense the future and demystify technology, which distinguishes his growing collection of works. *Home*

Attitude continues the tradition. Like a digital Rumpelstiltskin, John turns straw into gold. He spots technology trends, catalogs hundreds of products from the Internet of Things, to home operating systems, security systems, climate control, entertainment, geofencing, and pizza tracking. It's a piñata of shiny objects; a dizzying profusion of choices, for which you need this book, and a house doctor. Alexa, get me Dr. John!"

James G. Kollegger, CEO, Genesys Partners, Inc.

"For most of us, home automation is still a very new field. In *Home Attitude*, John Patrick shares his knowledge and experience from over 25 years in home automation. This book gives a comprehensive overview and many real-life examples. It will bring anyone's home attitude to the next level. It certainly did so for me."

Konrad Gulla, CEO/Founder, Keeeb, Inc.

Also by John R. Patrick

Election Attitude - How Internet Voting Leads to a Stronger Democracy (2016)

Net Attitude: What It Is, How to Get It, and Why It Is More Important Than Ever (2016)

Health Attitude: Unraveling and Solving the Complexities of Healthcare (2015)

Net Attitude: What It Is, How to Get It, and Why Your Company Can't Survive Without It (2001)

Dedication

This book is dedicated to my loving wife, Joanne. Her patience enables me to find time to work on books.

Preface

Many of my hobbies have been centered on technical things. It started with toys and blocks. They were much more than toys and blocks to me. They had practical purposes. The earliest I can recall building things was with Lincoln Logs. Then came Erector Sets. There were no Lego construction sets when I was a child and, if there had been, I would surely have been an enthusiastic builder. Chemistry sets, junior scientist kits, and amateur radio piqued my interest. In my adult years, the trend continued with GPS, digital cameras, personal digital assistants, personal computers, home automation, 3-D printing, virtual reality goggles, airplanes, motorcycles, and electric cars. Most new technologies arouse my curiosity.

In the late 1950's and early 1960's, when I was a teenager, building electronic kits from Heathkit was my favorite hobby. During the Heathkit era, which lasted from the late 1940's through the mid 1980's, it was possible to build a wide range of things including hi-fi/stereo systems, radios, TVs, radio and TV test equipment, amateur radio receivers and transmitters, and numerous consumer electronic devices.[1] My grandfather built a special workbench for me which included electrical receptacles for soldering irons and test equipment. Learning how to use a soldering gun properly continues to serve me well for various home automation projects and other tinkering.

Building a television set was more expensive than buying one already built, but the satisfaction and educational value were positive. Friends and neighbors would buy Heathkits and I would build them for them for free. The ultimate delight came from the last step of kit building. Turning on the power and seeing what were formerly hundreds of parts come to life was thrilling and a moment of pride. I can remember people asking me what kind of TV we had. I would proudly say, "We have a Heathkit TV, and I built it."

Heathkits were first marketed by mail order, with advertisements appearing in electronics and amateur radio magazines. I eagerly awaited new issues of these magazines to see if Heath had introduced any new kits. Even more exciting was the arrival of a new Heathkit catalog. The largest kit I ever built was the TX-1 "Apache" Ham Radio Transmitter. It had more than 1,600 parts to assemble. If Heath was still around, they surely would have home automation kits.

Home automation has been one of my hobbies for more than twenty-five years. In the early days, it was mostly lighting control using X10 technology. Pico Electronics of Glenrothes, Scotland, developed the X10 standard for home automation in 1975. Home power receptacles and switches compatible with the specifications of X10 could be turned on or off from a remote control. Although there are more advanced technologies available today, millions of X10 devices are in use around the world.

A researcher at IBM's Almaden, California laboratory wrote a software program which calculated the precise time of the daily sunrise and sunset based on the latitude and longitude of your location. This was significant because there was no GPS back then. Today, GPS is built in to cars, computers, smartphones, and watches. After I learned about the IBM program, I combined it with an X10 switch to turn my driveway floodlight on and off every day. At exactly sunset, a radio signal traversed its way across the electrical wiring of our home and turned on the light. At midnight, the light would turn off. This was an elementary example of home automation.

In the year 2000, when I began to think about retirement from IBM, I also started to plan a new, fully automated "smart home". My wife and I had weekly meetings with an architect to plan the physical aspects of the smart home. I had weekly meetings with an audio/video specialist who had an interest in home automation. We moved into the "smart home" at the end of March 2002. I have not written much about the home up to this point, somewhat out of modesty, but also not being sure how many people would be interested in what I had done. The builder was convinced home automation would make home resale impossible.

I ran into Steve Hamm, a journalist from *BusinessWeek*, in April 2006 in Rome at an IBM Business Leadership Forum. He learned about my home from a mutual friend, and he convinced me to open up about the subject. Steve visited our home the following September. I was confident he would write something thoughtful.

The article, "King of His Digital Castle", appeared in *BusinessWeek* on January 21, 2007.[2] Steve Hamm said, "At Patrick's suburban castle, almost everything that can be automated is, and it's all tied together in a network that acts like a central nervous system for a smart home. We're talking security, music, TVs, computers, home theater, lighting, heat and air conditioning, garage doors, window shades, propane, backup electrical generator, and outdoor spa."[3] I was embarrassed with the term castle, but Hamm got the key principles of home automation just right. There was a follow-on visit by *BusinessWeek TV*. Their filming ended up on *BusinessWeek Weekend* (carried by *ABC-TV*) on the Sunday after Christmas in 2007. During the period from 2002 to 2008 there were numerous newspaper, magazine, and TV stories about our smart home.[4] It was an expensive project, but I learned a lot from the experience of designing it.

Just like PCs and other consumer technologies, the cost of home automation has dropped significantly over the last decade. There remain issues of compatibility, complexity, ease of use, etc., but for the most part, home automation has become affordable, usable, and effective. Among the many benefits are improved efficiency for heating, air-conditioning, and lighting. Home automation can enhance the security of your home. It can provide remote access when you are away so you can adjust your thermostats or unlock your doors for a service technician. Entertainment systems can be automated to enhance your experience in using multiple devices and sources of content. You can forget about remembering to turn off lights.

Automating your home is much more than a hobby. It is an attitude, a home attitude. I have learned a lot about home automation over the past 25 years. Whether you are looking for a new hobby, seeking to enhance some aspect of your home, or just looking to experiment, I hope you will find *Home Attitude* enjoyable and useful.

Table of Contents

List of Figures

List of Tables

Introduction

I magine you are driving home from a winter vacation. You had put your departure time on your smartphone calendar, and at the appropriate time your smart home turned up the thermostats to ensure it would be warm when you arrived. As you approached within an eighth of a mile from your smart home, the garage door opened, and the security system was disarmed. The lights turned on in the mudroom, hall, and kitchen. When you walked into the kitchen, the counter top speaker made an announcement. "Welcome home. Today is Sunday, January 14, and the time is 9:15 PM. The temperature outside is 28 degrees and the humidity is 31.5. There are two messages on your home phone. Would you like to hear them now? Someone came to the front door last night, but did not give a name. I saved a picture to your smartphone. Have a nice evening". Classical music from KDFC then began to play throughout the smart home. If there had been a water leak while you were away, the water main would have automatically turned off, and you would have received a text message and a call would have been placed to your plumbing service. Any malfunction with the electrical, heating, or air conditioning systems would likewise have resulted in a text to you and a call to the appropriate service organization.

The opening paragraph is not science fiction. All these actions described for a smart home can be done today. The question for many consumers is how best to make their home a smart home. If reading the paragraph gives you an appetite for a smart home, you have the beginnings of a home attitude. Fortunately, there are several ways to make your home a smart home. If your budget is unlimited, you can hire experts, show them the paragraph above plus other goals you may have, and have them build it, test it, and hand you the keys. Another approach is to subscribe to a smart home service through your cable, telephone, or security service provider and select the offering which best matches what you want. The third approach is "do it yourself" (DIY). In *Home Attitude: Everything You Need to Know To Make Your Home Smart*, I will explain the pros and cons of the three approaches, but first some background.

A Brief History of Home Automation

Home automation is about controlling various devices in and around our homes by commands from our smartphones, our home automation systems, or simply by speaking to voice-activated devices nearby. The simplest definition of home automation is technology which enables a home to do things automatically, without human intervention.[5] Another term often used to describe the technology is the smart home. Ray Bradbury, the prolific science fiction writer, in his 1976 book *There Will Come Soft Rains*, said that in the future, homes would be interactive.[6] He envisioned homes would also be able to do things on their own, even after their human owners had died.

In the winter of 1969, the Neiman Marcus Christmas catalog offered a Honeywell Kitchen Computer.[7] It weighed 100 pounds and cost $10,000. The focus of the kitchen computer was to assist the chef in preparing meals by matching vegetables to other compatible foods. *Wired* reported not one of the Kitchen Computers were sold, but it was an early example of home automation.[8]

Pico Electronics in Glenrothes, Scotland, is an electronics company. In the mid 1970s, the company developed a remote control to enable a record player owner to change music tracks. The innovative project to develop the remote control was referred to internally as project X9.[9] If they could remotely change the track of a music record, they concluded they could remotely control other things. Project X10 resulted in a standard for how various devices could be controlled by sending a signal across the surface of the electrical wiring in a home. Beginning in 1978, X10 devices for controlling lamps and appliances were sold in Radio Shack electronics stores. It was at the time this technology debuted when I started experimenting with home automation.

In 1998, a prestigious group of technology companies, including Compaq, Honeywell, Intel, Microsoft, Mitsubishi Electric, and Philips Electronics, announced the Home API.[10] The group would develop an open industry specification touted as the key to creating a huge new market for home automation. The application programming interface (API) would ensure interoperability among various devices and software. The vision would mean everyone would collaborate and make things easy for the consumer. However, the Home API was limited to Windows PCs, and it never gained consumer adoption. Windows has been popular on PCs, but home automation apps today use smartphones. Microsoft has been insignificant in the mobile space compared to Apple's iOS and Google's Android platforms.

Microsoft subsequently launched HomeOS and then Lab of Things, both aimed at the home automation market. So far, neither Microsoft nor any other company have captured this market.[11] Apple is hoping to do with HomeKit what Microsoft was not able to do with Home API. Although Apple has a track record of making things easy to use, they may face some of the same challenges Microsoft faced, namely getting all the home automation vendors to work together to make things easy for the consumer.

Home Automation Today

One of the challenges in home automation is the lack of a single standard to make all the plugs, receptacles, switches, thermostats, fans, door openers, etc. compatible. This would enable a hobbyist or professional installer to easily develop a single automation system to control everything in a smart home. The problem is not the lack of a standard. The problem is there are many home automation standards including Bluetooth, X10, Z-Wave, Zigbee, Wi-Fi, and others. Home automation products supporting yet another communications standard may appear before the end of 2017. The new standard is called Thread. It is designed to make home automation devices look like Internet-connected devices.[12] I will discuss details about the most commonly used standards in subsequent chapters.

In addition to the multiple standards for the devices, there is the issue of which computing platform the home automation software runs on. For desktop users, there are three choices: Microsoft Windows, Apple's Mac OS, and Linux. Linux has many different versions. The most popular for consumers includes Arch Linux, Debian, Fedora, Gentoo, openSUSE, Slackware, and Ubuntu. For smartphone and tablet users, the dominant choices are Apple's iOS, Google's Android, and Microsoft's Windows 10 Mobile. Home automation software is available for all the desktop and mobile platforms.

A decade ago the cost to fully automate a home could be tens of thousands of dollars. Like other consumer electronic technologies, the cost has declined significantly. Depending on the degree of automation, it can be expensive, but the savings from improved energy efficiency can often offset the cost. For example, having your home automatically turning off your water main in the event of a water leak can avoid the cost of water damage which can be thousands of dollars for one incident. Despite the various challenges, home automation can enhance a home and save homeowners money.

The Internet of Things

In 2001, I predicted in *Net Attitude: What It Is, How to Get It, and Why Your Company Can't Survive Without It* there would be "millions of businesses, billions of people, and trillions of things connected to the Internet".[13] The first two predictions have come true. The total number of websites reached one billion in September 2014.[14] It is difficult to determine how many of these are businesses, but the website Internet Retailer estimates it ranges between 12 and 24 million.[15] According to Internet World Stats, as of November 15, 2015 there were 3.3 billion Internet users representing 46% of the world's population.[16] It is the third prediction, trillions of things, which may be more controversial. It was hard for many to believe in 2001 when I said it, but now, if you look at the details, the prediction becomes more believable.

The Internet of Things (IoT) is the interconnection of many things such as mobile phones, computers, and cars. In addition to being connected to each other through the Internet, they are connected to cloud services which allow them to share data with each other. The combination of these capabilities is called the IoT. Daniel Burrus, a leading technology forecaster and innovation expert, said, "Of all the technology trends taking place right now, perhaps the biggest one is the Internet of Things; it's the one that's going to give us the most disruption as well as the most opportunity over the next five years."[17]

Estimates of how many things will be part of the IoT range from 30 billion to a trillion. The number depends on how you define a thing. I define a thing as an electronic chip or device which can be connected to the Internet or be controlled by something connected to the Internet. From here on, I will refer to things as devices. They include smartphones, Wi-Fi TVs, most modern automobiles, and many home automation devices.

I have a range of devices connected to the home automation system in my smart home. These include:

❖ Amazon Echo artificial intelligence devices (Alexa)
❖ Apple TV
❖ Apple Watch
❖ Ceiling fans
❖ Door locks
❖ Door sensors
❖ Electrical switches and receptacles
❖ Electrical usage monitor
❖ iPhone and iPad
❖ iMac and MacBook Pro
❖ Keypads on walls
❖ Lights
❖ Light sensors
❖ Motion sensors
❖ Motorized solar shades
❖ Power monitor
❖ Remote controls
❖ Roku
❖ Security panel
❖ Sonos controller and speakers
❖ Switches
❖ Televisions
❖ Temperature and humidity sensors
❖ Tesla Model S
❖ Thermostats
❖ Video Projector
❖ Water leak sensors
❖ Water main solenoid
❖ Wi-Fi enabled televisions
❖ Window sensors

Like millions of households, my wife and I have computers, smartphones, and tablets, but home automation includes many more devices. The total of the devices in my home is approximately 200. If you included the multiple smart chips inside some of the devices, the number would exceed 300. For example, a smartphone with a camera and microphone could be considered as three devices. One of the devices in the list is my Tesla Model S, but it is much more than one device. With the Tesla iPhone app, I can honk the horn, turn on the lights, initiate charging of the car, open the charging port, turn on the air conditioning or heat, unlock the doors, and enable someone to drive the car away. Today's cars contain dozens of computers, some more than 50, which control ride handling, climate, entertainment, communications, and monitoring air pressure in tires.[18]

As of December 2015, the world population was 7.3 billion. If 25% of the population each had 275 devices, the total would exceed a half-trillion. Everyone does not have 275 devices today, but with nearly half of the world's population using the Internet, and cars and appliances containing more and more devices, the trend is clear. In addition to the billions of devices related to people, there are many more billions related to factories, hospitals, spacecraft, and research laboratories.

In *The Inevitable: Understanding the 12 Technological Forces That Will Shape Our Future*, author Kevin Kelly explained that in 2015, five quintillion (10 to the power of 18) transistors were embedded into objects other than computers. He said, "Very soon, most manufactured items, from shoes to cans of soup, will contain a small sliver of dim intelligence."[19] My prediction of trillions of things connected to the Internet is not as far-fetched as it seemed when I wrote about it in 2001. Although growing fast, the Internet of Things is still in its infancy. Home automation technology is a subset of the Internet of Things. A home becomes a smart home by using home automation technology.

The focus of *Home Attitude* is to help consumers create and use smart homes. The book is organized into two parts. Part 1 describes what home automation is and what it can do for you. The benefits of home automation discussed include the following areas: lighting control, home entertainment, home security, energy efficiency, weather monitoring, and convenience. It concludes with a chapter about security and privacy.

Part 2 is about how home automation works. Without using complicated technical terms, the chapters in this part describe the principles of home automation and the basic home automation components including hubs and networks, devices and plugins, actions and scenes, schedules and triggers, and user interfaces.

The last chapter will help you with the choices in getting home automation to work best for you. It describes the pros and cons of hiring a professional to design and implement your home automation system versus doing it yourself as a hob-byist. The other choice explored is whether to have your home automation system operate from a computer in your home or using a cloud based service offered by a home security or telecommunications company. The chapter concludes with suggested next steps.

Part 1
Home Automation What It Is and What It Can Do For You

CHAPTER 1

Home Automation Basics

The Oxford English Dictionary defines automation as the action of intro-ducing automatic equipment or devices into a process or facility.[20] The facility in home automation is the home, and the automation is when something in the home happens automatically. If you flip the kitchen light switch to turn on the ceiling lights, the switch flip is not automation. If it becomes dusk and your outdoor lights turn on automatically, this would be a simple example of automation. If it is dinner time and you push a button labeled "dinner" on your smartphone and multiple actions take place automatically, this would be a more significant example of automation. For example, pressing the dinner button could cause the lights over the stove and kitchen sink to go out, the under-counter lights to dim to 50%, the chandelier above the kitchen table light to turn up to 40%, and your music system to begin to play an easy listening station from Pandora at a volume level of 3. The kitchen dinner scene just described would be an example of home automation you could activate when you are at home. Some home automa-tion takes place when you are away from home. For example, you could program your exterior lighting to turn on at 15 minutes after sunset and then turn off at sunrise even though you may be on vacation 1,000 miles away.

In the September 2015, Digital Edition of *PC Magazine*, Alex Solon wrote about home automation products for the home. He described 12 devices for each of five parts of a home: living room, bathroom, bedroom, kitchen, and back yard.[21] One of the most interesting devices he presented was the iCPOOCH. The innovative device lets you make a video call to your pet. The iCPOOCH unit is about the size of a large toaster. It sits on the floor and has a video display and speaker so your pet can see and hear you. While you are interacting with your pet, you can tap on the smartphone app and the iCPOOCH unit opens a small trap door and dispenses a treat. Solon said, "It's a good way to feel connected to your dog or cat, even at those times [when] you're far from home."

Although connecting with a pet may be nice for some people, most home automation is based on connecting with a plethora of devices such as light switches, smart lightbulbs, door locks, thermostats, entertainment systems, and various kinds of sensors. One of the challenges associated with home automation is the proprietary nature of many of the devices. For example, if you buy a Nest thermostat, you use a Nest app on your smartphone to adjust the settings of the thermostat. If you have a Liftmaster garage door opener, you can connect and open the door remotely using the Liftmaster MyQ app on your smartphone. The Kwikset Kevo Smart Lock can be operated using an eKey in the Kwikset smartphone app. If you want to control your Hunter ceiling fan, you use the Hunter Fan SimEconnect™ Fan Control smartphone app.

These are just a few examples. There are many more. The proliferation of devices and apps can become very confusing. Downloading and using a multitude of apps can offset the productivity gains possible with home automation. A home automation system built with a home attitude can enable you to have a smart home with one app which can integrate and control all the devices. Fortunately, the technology exists to do exactly that. In the chapters to follow, I will describe how an integrated system works to make your home a smart home.

Getting a Smart Home

There are many systems, services, products, and vendors to choose from to create your smart home. The choices fall into three categories: subscription based smart home services, professionally built custom systems, and do it yourself hobbyist approaches. Each choice has pros and cons.

Subscription based smart home services are typically offered by telecommunications providers such as AT&T and Comcast, and security monitoring companies such as ADT, Safe Streets USA, and Frontpoint. These companies are already providing phone or security services for the home, and it is a natural extension for them to extend their services to include home automation.

Professionally built custom systems are available from companies which operate as systems integrators. They typically are business partners of companies such as Control4, Crestron, Leviton, Lutron, and Savant, all of which make home automation products. The integrators either offer security monitoring or partner with a company which provides it.

The third choice is Do It Yourself (DIY). A DIY approach can include all or just a part of the smart home. In some cases the DIY consumer will add a few features such as lighting control to a home security system provided by someone else. Some DIY consumers start from scratch and create an entire smart home with all the available bells and whistles one can imagine. I will discuss the pros and cons of all three approaches in more detail in the last chapter.

The Basics

Although there are many approaches, options, and features available when creating a smart home, there are some basic components which are quite similar. I call these home automation basics. They include devices, hubs, schedules, action groups, triggers, scenes, and user interfaces. In the remainder of this section, I will summarize what each basic is about, and then explain them in more detail in Part 2 Home Automation: How It Works and How to Get It.

Devices

Merriam-Webster.com defines the word device as "A piece of equipment or a mechanism designed to serve a special purpose or perform a special function".[22] In the world of home automation, which makes your home a smart home, the devices must be smart devices. For example, iPhones and Android phones are smart devices. In fact, they are exceptionally smart. In *Health Attitude: Unraveling and Solving the Complexities of Healthcare*, I called them personal supercomputers (PSCs).[23] From here on, when I use the word device, I mean a smart device. Smart home devices have a computer chip in them which makes them smart and enables them to communicate, either via a wire or wirelessly. The devices can communicate with other devices or with a home automation hub, which acts like a central point of contact for all the smart devices.

Devices come in many sizes and shapes, and can perform many different functions. A simple example is a smart lightbulb. The bulb has a chip in it. When it receives a command from a device or hub, it will turn itself on. A different command causes it turn off. Some more sophisticated lightbulbs can also respond to commands to dim or brighten. Some even smarter bulbs can change the color of the light they emit.

Computer devices in a smart home can include desktops, laptops, smartphones, tablets, and watches. Devices for lighting include switches, keypads, dimmers, bulbs, and receptacles. Entertainment devices include audio speakers, flat panel displays and TVs, remote controls, and devices for streaming of audio and video content. Sensor devices can detect sunlight, ultraviolet temperature, humidity, water leaks, motion, and vibration. Devices with motors in them include ceiling fans and solar shades. Smart thermostats can respond to commands to change heat and cooling set points, HVAC mode, and fan mode. Specialized devices can open or close a water main valve, start an irrigation system, or communicate with and control other devices.

Hubs

Every smart home has at least one hub. A hub can be a specialized device or a desktop, laptop, smartphone, tablet, or watch which has a home automation software app. Hubs act as a central control point for other devices. A hub can connect with devices via a wire cable or wirelessly, and issue commands for the connected devices to follow. A hub can be as simple as a single purpose hub to control a group of lights, or it can be a very sophisticated hub which can control everything in the smart home including all the devices mentioned earlier. Hubs have capabilities very specific to home automation which enables a home to become a smart home. At a very basic level, a hub tells a device what to do and when to do it. To make this possible, hubs have special capabilities called methods. The methods include schedules, action groups, triggers, and scenes. An additional capability of a hub is a user interface which allows users to interact with the various methods.

Schedules

Schedules tell devices what to do and when to do it. For example, a hub may have a simple schedule to turn the porch light on every day at 7 PM and then turn it off at midnight. The time can be 30 minutes before or after sunrise or sunset. Times can be plus or minus a random number of minutes within a certain range so the smart home can appear to be occupied even though there is nobody home. A schedule can be executed automatically every day, certain days of the week or days of the month, or variations such as the first and third Tuesday of every other month. A summer schedule could start on June 15 and end at the end of August.

Schedules can be customized to only execute under certain conditions. For example, a ceiling fan in the office could be scheduled to turn on to medium speed every week day, but only if the room temperature is above 70 degrees. A garage door opener could be scheduled to close every day at sunset, but only if it is open. A user might schedule SiriusXM Escape Easy Listening music to play at cocktail time on weekend days. The possibilities for smart home schedules are endless

Action groups

Merriam-Webster.com defines an action as, "a thing done".[24] In smart home context, a thing done might be turning on a light, but actions can do much more than turning a device on or off. Actions can include changing the heat set point of a thermostat, playing a music source, changing the music volume, or announcing the time of day or weather forecast through an audio device (speaker). An action can also be to set the value of a variable. For example, the hub might contain a variable called home_status. The value could be set to "home" or "away", and then be used to condition certain actions or schedules. Another action could be a notification, which can be by text, email, or home audio. The message could be about a leak detected in the basement where attention is required.

An action group is a group of actions which can be executed together. For example, the hub might contain a Good Night action group which includes actions for turning off smart home lights, lowering the volume of music for ten minutes and then turning it off, setting thermostats to a night setting, and locking the doors.

Triggers

Once again, Merriam-Webster.com offers a simple definition. It defines a trigger as, "something that acts like a mechanical trigger in initiating a process or reaction". In simpler terms, a trigger makes something happen. A smart home example of a simple trigger is a button push. The push of a button on a smart switch or a smart keypad could signal to the hub to make something happen. The something could be as simple as turning a light on or off or as comprehensive as executing a Good Night action group as described above.

A trigger can be quite sophisticated. A power failure in a smart home can trigger an email to alert you. A water leak on the floor can trigger an action group which includes shutting off the water main to minimize damage and then sending an email to you or your landlord. Walking in your front door can be detected by a motion sensor and result in a trigger which changes the variable home_status from "away" to "home". The trigger could also initiate an action group which turns on certain lights, adjusts the thermostats, and plays your favorite music.

Scenes

Scenes are designed to simplify the automation of the smart home. For example, if you want to automate lighting in the kitchen at dinner time, you could create an action group containing an action for each light. The kitchen might have counter lights, ceiling lights, kitchen table lights, etc. Each action in the action group would turn one of the lights on or off. An alternative method is to create a scene in the hub which includes all the kitchen lights. With a scene in place, automating the lights is a simple matter of one action which turns the scene on or off. Hubs, schedules, action groups, and scenes all work together to make your home a smart home.

User interfaces

Hubs, schedules, action groups, and scenes can only work together if there is an easy to use way for the homeowner to interact with them. Homeowners can issue commands directly by pushing buttons on a device such as a thermostat, door lock, light switch, or keypad such as shown in figure 1. However, to initiate an action group, scene, trigger, or schedule, the homeowner must interact with a hub or instruct the hub to act on its own under certain conditions or schedule. The interfaces available depend on what kind of home automation solution you have, but most solutions have both a web page and a mobile app. In both cases the interface presents a menu or buttons the user can select such as Good Morning, Lights Out, Open Garage Door, etc.

Some hubs can accept an email. For example, a homeowner could send an email to myhub@mysmarthouse.com with Security Status in the subject line. The

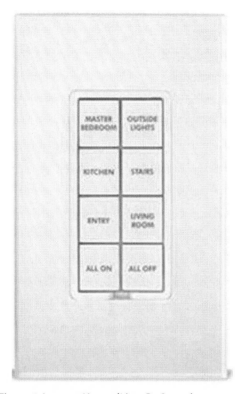

Figure 1. Insteon KeypadLinc. By Smarthome.com.

hub would know to send an email to the homeowner listing the open or closed status of all the doors and windows of your smart home.

A more natural way to communicate with a hub is by voice. Talking to your home, sometimes called voice activation, and asking to turn on a light or put down the shades is not a new idea. I tried it out quite a few years ago, but found it to be very unreliable. The microphone and translation technology was primitive. I would find myself yelling at my home and often getting the wrong action.

Today, controlling things by voice has become very sophisticated with the advent of highly accurate voice recognition and the use of artificial intelligence (AI) to interpret what you want done. As of this writing, Amazon has a significant lead over Apple and others with its Echo technology. If I say, "Alexa, turn on the office lights", it works 100% of the time. Echo does not interface will all hubs, but I am confident it will in the very near future. Without a doubt, Amazon's AI will soon enable you to say, "Alexa, I am cold". The Echo will then communicate with the hub, determine what room you are in by motion sensing, read the current temperature, and raise the heat set point on the smart thermostat.

We have now covered the basics. I hope this chapter has given you a good introduction to the potential of home automation and some ideas on how to make your home a smart home. In the remaining chapters of part 1, I will discuss more about what home automation can do for you if you have a home attitude. The applications I will discuss include lighting control, home entertainment, home security, home utilities, weather monitoring, and convenience.

CHAPTER 2
Lighting Control

Abraham Lincoln read from the light of a fireplace, and then would take a book to bed and read by the light of a candle.[25] About 15 years after Lincoln's assassination, a new source of light for reading was introduced. It was called the incandescent lightbulb. Incandescence occurs when a wire filament with electricity passing through it heats to a very high temperature, which makes it glow with visible light. The bulb is needed to protect the filament from oxidizing and burning out. Incandescent bulbs are available in a wide variety of sizes and light output. Although incandescent bulbs have served us quite well for many years, they are highly inefficient. On average, incandescent bulbs convert about 2.2% of the energy they use into visible light, with the remaining energy being converted to heat.[26] Another disadvantage of incandescent bulbs is their lifetime. A typical home lightbulb lasts 1,000 hours compared to 10,000 hours for compact fluorescents, and 30,000 hours for LEDs.[27]

Research on light emitting diode (LED) semiconductors dates to 1907, but in recent years the market price and market share have changed dramatically. Since 2008, the price of an LED lightbulb has dropped 90 percent.[28] As of March 2017, 10 watt LEDs were available online for less than $3. Initially very expensive and only available with a few specifications, LEDs today are available in many sizes, shapes, light color, and for use in a wide range of applications. LED market share has gone from 1% in 2010 to 28% in 2015, and is projected to be 95% by 2025.[29]

Even though the availability of LED lighting has reduced the cost of illumination dramatically, most people do not want to leave the lights in their home on all the time. Home automation can enable a smart home to have the right lights on at the right time. When you push the Good Night button, most of your lights turn off. Some turn off instantly, while others may go out gradually. No more worries about forgetting to turn off a light in the garage or basement. Exterior lights remain on until a desired number of minutes before or after sunrise when they automatically turn off.

When you push the Good Morning button, lights begin to turn on. The nightstand table lamp may gradually ramp up over a period of ten seconds to give your eyes a chance to adjust. The bedroom and bathroom lights turn on. Five minutes later, the nightstand table lamp turns off. The following sections describe various ways you can control your smart home lights.

Scenes

As a homeowner, you can turn a light on or off by flipping a switch or pulling a chain. Some years ago, my wife and I were looking at a home for potential purchase. The thing I remember most vividly about this particular home is the switch plates in the kitchen and adjoining dining area. There were 25 switches, grouped in sets of two, four, six, or eight. Remembering which switch does what must be a challenge for the owners.

In a smart home, you have many other options for how to control lighting. One method is the use of scenes stored in a home automation hub. For example, you can have a Cooking Scene and a Dining scene for your kitchen lighting. Scenes are composed of groups of lights in a room or area. Kitchen and dining areas may have quite a few lights in rooms or areas of rooms. See table 1 for an example of

Table 1. Kitchen Lighting Groups

Lighting Group	No. of Lights
Over the Sink	2
Over Cooktop Stove	3
Over Dining Table	1
Under Counter Left	4
Under Counter Right	4
Main Ceiling Lights	4

a kitchen and adjoining dining area. There is a total of 18 lights over the sink and cooktop stove, over the dining area table, under the counters, and in the ceiling.

The scenes consist of a collection of lights from the lighting groups and corresponding specific levels of brightness for each of the groups. When it is time to cook, you turn on the Kitchen Cooking Scene and all the lights in the scene are dimmed or brightened to exactly the level you want. See table 2 as an example of the six groups of lights controlled by a single scene. When it is time to sit down to dine, you turn on the Kitchen Dining Scene. The six groups of lights change to different pre-defined brightness levels as shown in table 3. The scene can also flash the lights on and off a few times in the TV room to let the kids know dinner is ready.

Table 2. Kitchen Cooking Scene

Lighting Group	Brightness
Over the Sink	100%
Over Cooktop Stove	100%
Over Dining Table	20%
Under Counter Left	50%
Under Counter Right	50%
Main Ceiling Lights	80%

Table 3. Kitchen Dining Scene

Lighting Group	Brightness
Over the Sink	0%
Over Cooktop Stove	0%
Over Dining Table	50%
Under Counter Left	20%
Under Counter Right	20%
Main Ceiling Lights	10%

Scenes can create a mood. After the Kitchen Dining Scene is turned off, it might be time for reading and relaxation in another room. An eclectic scene could use various colors of LED lighting which fade from one color to another every so many minutes. Lighting scenes can be combined with other actions. For example, with the touch of a button, the lighting scene can be accompanied by your favorite music consistent with the room, time of day, or your mood.

Scenes are limited only by the imagination. Grant Clauser, a writer and editor at *The Wirecutter*, has been writing about home automation for 15 years. In a recent article in *Electronic Home*, he said,

> Most people think of their home lights as things that get turned on, off or dimmed as needed, but not as a design and lifestyle element that can be programmed to fit a family's moods, activities or schedule. That's what smart lighting control systems do. They can light your home and yard, but they can also turn a dull standard glow into an integrated part of the home environment.[30]

The following sections paraphrase and expand on some basic scenes which Grant described in "Smart Lighting Scenes for Home Automation and Lighting Control Systems".[31]

Morning

The morning scene, which you might call Good Morning or Wakeup, is the most important because it sets the tone for your day. Your bedroom table light can come on full blast to jolt you awake, or gradually, over a period of 30 seconds, to ease you into the day. The morning scene can also turn on a path of lights from the bedroom to the hall, down the steps and into a nicely lit kitchen. It can also turn off any outside lights or dimmed lights you chose to leave on. Your scene can include other actions such as announcing the date, time, and weather, and turning on the morning news from your favorite TV or streamed audio station.

Night

A night scene, which you might call Good Night, can turn off all the lights in the entire home at bedtime. Like the Dining scene described earlier, you can decide which lights go all the way off, which stay on at full brightness, and which should be left on but dimmed. For example, a driveway light might stay on at full brightness until midnight and hall lights might be dimmed to make it easy for family members to see their way to the bathroom.

The Good Night scene can be combined with other actions. You can have the thermostat's set points automatically adjusted for comfortable but economic sleeping. After all locks are confirmed to be set, the security system can be armed. Music can be turned off in all rooms except the bedroom, where it will play for a half hour and then fade away along with the bedside table lamp.

Party

A party scene can create whatever mood you like. The simplest would dim the lights in the family room for conversation or dancing, and illuminate the kitchen counter top where the food is or over the bar where the drinks are. Party scenes can be turbocharged with bright and colorful smart bulbs. Smart color bulbs can glow with any color of the rainbow and set to blink, fade in and out, and change colors. You can take the party to a higher level with rotating color disco stage lights.

Movie Time

If you are a videoholic, every night may be movie night at home. The Movie Time scene can turn off all lights which may be distracting from the TV or video screen. The scene can also include some dimmed lighting near the fridge or bathroom. You can be more selective by having three movie scenes: Movie Start, Intermission, and Movie Over. The latter two might include some of your favorite music.

In my home, I have a movie scene which turns the lights on in a path to the movie theatre room. The video projector rises from a hidden space in a counter and a large screen descends from a hidden slot in the ceiling. As the projector bulb warms up, the light path entering the theatre and the ceiling chandelier gradually dim. When the projector is at full brightness, the lights are all dark.

Vacation

A Vacation or Away scene can play multiple roles. Turning off all unnecessary lights can save electricity. During an extended vacation, the scene can be accompanied by reduced heating or air conditioning settings. The water supply can be turned off to avoid costly damage from leaks.

In addition to arming the security system, the lighting can play a security role during vacation. The lighting scene can be programmed to mimic the way the lights would operate if you were home. The lights can come on in various rooms during the time they would normally, but within randomized time windows. For example, downstairs lights can come on at a random time between sunset and 30 minutes after sunset. They can then go off at a random time between 10:00 and 10:30. The upstairs lights would come on at the same time. In other words, the lighting would appear as it usually does but not exactly. The random scene can be a helpful way to deter criminals.

Reading and Writing

A reading scene in my home office is my favorite. I watch an occasional movie, but I prefer reading and writing. Most of my news consumption is from reading online. The Read scene turns on the sconces closest to my chair, turns off the room's other lights, and sets music streaming to one of my favorite classical music sources.

Paths

An important use of scenes is for the creation of lighting paths inside your smart home. For example, you can have a path of lights from your kitchen to your family room. When you turn the scene on, the hall lights turn on to illuminate your walk from the kitchen into the family room. Thirty seconds after you arrive in the family room, the hallway and kitchen lights turn off. Paths can help you have light where you need it and not waste any electricity for lighting where you do not need it.

Specialized Scenes

Scenes can be designed for any purpose. Ron Roslasky, owner of Intra Home Systems in Pompton Plains, NJ, created a lighting scene for his Labrador Retriever.[32] When it is time for a night trip to the backyard, he presses the DOG WALK button on a keypad next to his bed. The button launches a pathway scene. Lights come on from the bedroom and downstairs to the mudroom. The backyard spotlights illuminate the yard. When Rover has completed his evening duties, Ron heads back to bed and presses the DOG WALK button again. This time, the scene turns off in reverse. The flood lights go off first, followed by the lights from the mudroom and up the stairs. The bedroom lights go out after a 15 second pause.

Roslasky also has a clever WELCOME HOME scene.[33] As the car enters the half-mile-long driveway, sensors buried underneath the macadam send a signal to the smart home, which then turns on the landscape lights. Rather than turning all the lights on at the same time, they turn on sequentially. As the car drives up the driveway, the lights turn on every 200 feet.

If This Then That

A little more than five years ago, a couple of tech entrepreneurs in San Francisco got the idea to develop something new. The idea was to create a free web-based service which allows users to create chains of simple conditional statements, called applets, which are triggered based on changes to other web services such as Gmail, Facebook, or Weather Underground. They called the idea IFTTT, an abbreviation of "If This Then That". It sounds complicated, but you will find it simple after I provide some examples.

A simple example is about weather. If wunderground.com shows a forecast of rain where you live, then you would receive an email. A variation would be if the forecast is rain, then the forecast would be placed on your Google calendar. Following are a few more examples. If the President signs a bill into law, you would receive an email about it. If the New York Times has an article about the Affordable Care Act, then the article would be sent to your inbox. If you are within 10 miles of your home, the thermostat would be changed to your desired setting. If you receive an email with an attachment, then it would be saved to your Dropbox.

There are millions of possibilities and the idea is growing rapidly. You can browse through available applets at IFTTT.com. Applets can be a convenient way to gain daily productivity, but the city of Louisville, Kentucky had a bigger idea. They have created a family of applets, called Smart Louisville, to benefit asthma sufferers. One of the Smart Louisville applets will retrieve the Air Quality Index from Louisville Metro Government's Air Pollution Control District. The applet works with a Philips Hue smart lightbulb which can change color. If the air quality is good, the bulb turns green. If it is moderate, the bulb turns yellow. An unhealthy air quality turns the bulb red, and a hazardous condition turns the bulb dark purple.

Domino's delivers more than 1 million pizzas a day worldwide.[34] The public company is 97% franchise-owned, operates 13,800 stores in more than 85 countries around the world, and has more than 260,000 employees worldwide.[35] Global retail sales for 2016 were $10.9 billion.[36]

In 2008, Domino's launched an innovative delivery tracker. It is easy to order pizza online in up to 34 million variations.[37] The hard part is the waiting and wondering. The Domino's online delivery tracker lets you watch your order go from prep, to bake, all the way through to delivery. In May 2017, IFTTT announced the ability to create applets which work with the tracker.

The simplest example would be to get a notification text or email when your pizza is on its way. If you apply a home attitude, you can do much more. IFTTT said, "You can use smart home devices to come up with your own creative Pizza Party Imminent warning system. When your IFTTT applet gets the Order Received signal, you can cause your smart home to turn a Philips Hue smart lightbulb on in various colors. Yellow could mean your order is being prepared, red indicates it is in the oven, and green means your order is Ready for Pickup or Out for Delivery. You could also ask your home audio system to play the Pizza Power song by Teenage Mutant Ninja Turtles. If the pizza is being delivered, you could have your front lights blink to help the driver come to the right place. Domino's has built a few Applets to give you more ideas.[38]

Geofencing

A clever technique to turn a scene on or off uses Geofencing. A geofence is a virtual perimeter for a real-world geographic area. For example, you can use an app such as Locative to define a geofence around your home. The center of the virtual perimeter geofence around your home is the latitude and longitude of the location of your home. You can specify the radius to be whatever you want, say 100 feet. As in IFTTT, if you walk or drive outside of your geofence, then your Away scene is activated, turning off lights, arming security, and setting thermostats. When you return, and enter the geofence, your Home scene is activated, turning on lights, disarming security, and setting thermostats. The nice feature when using Geofencing is you don't have to remember to activate your scenes.

Schedules

Scenes can be activated by pushing a button on a remote or on the wall. They can also be activated by use of a geofence or other forms of IFTTT. Another way to activate scenes is with schedules. The concept is simple. You establish a time for a scene to turn on and a time for it to turn off. However, you can make schedules much more sophisticated. I have a scene which turns terrace lights on 15 minutes after sunset and then turns them off four hours later. The schedule is only followed if I am home.

A schedule can be hourly, daily, every other day, only certain days, weekly, monthly, or yearly. The times selected can be random by plus or minus a specified number of minutes from a certain time. They can activate only in the summer or winter or between specified dates. They can be active only if certain conditions are met, such as being home or only if it is dark.

After using lighting scenes for a while, you may decide to tweak which lights are included and what brightness levels you want to set. You can also override scene settings, if you want a change. If you want a room a bit brighter, you can simply go to the wall switch dimmer or use a smartphone app to adjust the light to whatever level you want. Schedules can be designed to follow your way of life. If your preferences change, you can change your schedules.

Security

In the section about a Vacation scene, I described the randomizing of lighting to make someone surveilling your smart home to think your home was occupied. However, it is possible someone could be successful in breaking into a smart home undetected and without tripping the security monitoring. In such a case, a motion sensor could trigger an Alarm scene which would turn all lights on to full brightness and trigger the security system to report the intrusion. This is a good example of how a home attitude and a smart home can add value. Security systems are typically standalone isolated systems. In a smart home, the security system and lighting scenes are integrated to enhance the home's security.

Summary

Lighting scenes are practical and can enhance living in your home. They offer unlimited creativity, and you can refine them to mirror your preferences. Lighting scenes can strengthen your home security while reducing your cost of electricity. Another practical use of home automation is the control of your home entertainment systems.

CHAPTER 3
Home Entertainment

W hen I was growing up, watching a movie or listening to some music was as simple as walking downtown to the movie theatre or turning on the radio or TV. Today we have a multitude of options. There are many forms of entertainment available in the home without braving the weather or traffic to drive somewhere. In this chapter I will highlight some of the more popular forms of home entertainment and describe how a home attitude can automate and enhance the entertainment experience.

Music Makes the World Go Around

George Rachiotis, a music journalist, said "Music is the greatest creation of man."[39] In his blog post, "The Importance of Music in Our Daily Lives", Rachiotis said that music draws people together to uplift them emotionally, helps humans to express themselves with ease, and has a therapeutic effect in our lives.[40] Music has always been a part of my life, and I listen to it during most of the hours of every day.

MP3

Fortunately, there are many ways to listen to music. In days past, people listened to music from vinyl records, various tape formats, and more recently CD/DVD players. In 2014, revenues from digital music services matched those from physical format sales for the first time, and analog music continues to fade as digital music pushes it aside.[41] Digital music is available in many different formats, but the MPEG Layer-3 (MP3) format is the most popular format for downloading and storing music.

Brothers Shawn and John Fanning, along with Sean Parker, launched an Internet service called Napster in 1999. The technology, called peer-to-peer (P2P) file sharing, was easy to use and was specifically designed for sharing digital music files, in the MP3 format, across the Internet. The recording industry took Napster to court, and was successful in convincing a judge the Internet service was designed and was being used to steal music. The judge shut down the service in 2001. See appendix A for a description of MP3 and how it works.

MP3 has changed how people think about digital music, whether they are consumers, artists, producers, or broadcasters. The recording industry, understandably, reacted very negatively toward Napster and other music sharing programs on the Internet. They viewed the programs as methods for stealing, and have continued aggressively to shut down websites they view as contributing to theft in any way.

Alternatives to the sharing programs emerged in the late 1990s. A patent for the first digital audio player was filed in 1981 and issued in 1985. Over the ten years to follow, several business and technical models were developed, but none gained market traction. The first MP3 player was developed by SaeHan Information Systems in 1997. The Rio PMP300, from Diamond Multimedia, was introduced in September 1998. The Rio was successful and spurred interest in digital music.[42] The recording industry filed a lawsuit alleging the device abetted illegal music copying, but Diamond won a legal victory, and MP3 players were ruled to be legal.[43]

The destiny of MP3 digital music changed dramatically in 1998 when Bill Kincaid developed SoundJam MP, a software program which could play MP3 music. The product was released by Casady & Greene in 1998.[44] In 2000, the software was purchased by Apple and renamed iTunes.[45] On November 10, 2001, Apple released the hardware component of iTunes, which it named iPod.[46] As of 2016, Apple had sold nearly 400 million iPods.[47] Over a span of 15 years, iTunes evolved into a sophisticated multimedia content manager, hardware synchronization manager, and e-commerce platform.

MP3 players still exist, but the trend is shifting rapidly to streaming music services. As of September 2016, 100 million people were paying to listen to music on their computer, tablet, or smartphone from streaming music services such as Amazon Prime Music, Apple Music, Calm Music, Google Play Music, Pandora, SiriusXM, Slacker, Spotify, and many more. With subscription prices ranging from

$5 to $10 per month, the streaming music business has already grown to between 6 and 12 billion dollars per year.

Podcasts

Podcasting, first known as audio blogging, has its roots dating back to the 1980s.[48] The name podcast comes from a combination of iPod and broadcast. A podcast is like a radio program. Anyone can be a podcaster and millions of people listen to a huge variety of podcasts. Dozens of websites provide podcast distribution at little or no cost to the producer or listener.

A widely-used site for podcasts is Stitcher.com. The site offers fresh episodes of podcasts streamed directly to your smartphone, tablet, or desktop. The site offers more than 65,000 radio shows and podcasts. Popular shows include NPR's Fresh Air, Adam Carolla, Rush Limbaugh, Rachel Maddow, This American Life, Newsy, Freakonomics Radio, USA TODAY Talking Tech, Planet Money, PBS News Hour, and the Tim Ferriss Show. You can also listen to podcasts on your Amazon Echo or while driving a car with Apple's CarPlay.

Sound Systems

Millions of consumers are enjoying streamed music, news, or podcasts in their cars, on airplanes, bicycles, and motorcycles. They listen while hiking, biking, driving, running, or walking. There are many choices for listening devices ranging from ear buds for less than $10 to HIFIMAN HE1000 V2 Over Ear Planar Magnetic Headphones for $2,999. The HIFIMAN claims to have an "Advanced Asymmetrical Magnetic Circuit", which offers "perfect reproduction of live music".

Listening to audio on the go is great for many, but whether an on the go listener or not, many like to listen to their audio at home. There are numerous high-quality home audio systems to choose from including Bose, Harman Kardon, Korus, Sonos, or Sony. A sound system can be as simple as a bookshelf speaker or "whole home" systems with speakers in every room and weatherproof speakers outside. Most systems connect to the Internet using your home Wi-Fi and stream music or podcasts programs to any or all your speakers.

A smartphone can act as the remote control to connect to your sound system, select the desired music, and set the volume level. You can choose what music you want to play in which room of your home with each at different volumes. Some systems, such as Sonos, have a music balancing system which tunes the sound characteristics of each speaker to the room size, layout, décor, speaker placement, and any other acoustic factors which can impact sound quality.

You can apply a home attitude to your music by incorporating home automation to play a complementary role to your music system. Either at a preset time or at the touch of a button on a mini-remote control on your night table, your home automation system can start your day. You can wake up to your favorite music. My home automation system is programmed to first determine the day of the week. If it Sunday, the system selects Sunday Baroque on WSHU-FM. If it is not Sunday, the system randomly selects a streaming music channel from a list of a dozen of my favorite sources. The music starts at a volume of 5 (out of 100). It then gradually increases to 20. An announcement feature converts text to speech and plays it through my bedroom speaker. For example, it can say, "Good morning. It is seven AM. The temperature outside is 72 degrees, and the forecast calls for a high of 80 with a mostly sunny sky. Your first appointment is a conference call at eight-thirty with Bill Smith." Then, the music resumes.

The Good Morning scene can do more than the announcement and starting the music. It turns on a light path to the exercise room and turns off the lights behind me after a short delay. The scene sets the music level in each room to a low level so as not to disturb anyone while I am exercising. After a workout on the machines, it is time to turn off the news. Now the music level in the exercise room increases to a good listening level while finishing up with some stretches and floor exercises. Then it is time to go upstairs to continue the day. At the top of the stairs is a keypad. Pressing the Exercise Complete button pauses the downstairs music and causes the downstairs lights to time out after I have arrived at the main floor.

In summary, listening to audio is a very easy home entertainment option to enjoy. It can be extended to wherever you may be inside or outside the home. The choices of audio sources are broad and the pricing ranges from free to "all you can eat" fairly priced subscription services. Consumers can get what they want, when they want it, on any device, at a fair price. Video is quite a different story.

Let's Go to the Movies

In the mid 1950s, I walked across the street to visit a friend who had just gotten a brand-new television set. Getting a TV set was very special at the time. As of 1953, only 50% of American homes had one.[49] The video was displayed on a picture tube. The tube was sealed and contained a vacuum like a lightbulb. The picture was black and white. Color TV sets were just becoming available, but they were quite expensive. The visible viewing area of my friend's picture tube was five inches by seven inches. The TV set at my home was closer to 15 inches diagonally. My friend turned on the TV and we sat on the floor watching, quite impressed. I don't recall what we watched, but I remember part of what we were watching was not visible at the edges. I explained the problem would not exist on our TV because it had a bigger screen. My friend's mother heard the comment and rightly corrected me, explaining to me the video content on both TVs was the same even if the sizes of the TVs were different. I could not have imagined a future where an HD color movie could be watched on a four-inch iPhone or a home theatre with a 102" screen.

Back in the '50s, it was an exciting experience to see a new movie, such as *Lady and the Tramp*[50], at the local theatre downtown. My mother would give each of my two brothers and I a one-dollar bill. We would walk downtown to the barber shop for a haircut. The cost was 50 cents. We would then walk across the street to the movie theatre and pay 25 cents admission. We would buy popcorn and a Coke with the remaining 25 cents. Aside from 60 years of inflation in the price of movie tickets and popcorn, there are similarities to today.

Amazon, Apple, Netflix, and others would like us to enjoy movies streamed to our various devices whenever we want. Unfortunately, there is a major roadblock to make this possible: Hollywood. Hollywood distributes new movies through what is called a windowing system. During the first window, typically 90 days, a new movie is only available at movie theatres, which pay Hollywood fees to have the exclusivity. It ensures revenues for the theatres by preventing consumers from bypassing the movie theatres and relying on their own home theatres. A second window would include the theatres plus video on demand, where cable companies pay Hollywood to allow them to make the movie available for its subscribers. A third window would include video on demand plus availability on DVD. There can be up to a half-dozen windows with various combinations of pricing and time restrictions.

The sanctity of Hollywood's windowing system has remained intact for many years. The discussions and negotiations on the details take place between Hollywood producers, TV stations, cable companies, and Internet streaming services. The consumer is not at the table. We may be willing to pay to see a new movie at home on day one, but Hollywood prefers to preserve the movie theatre model because it is more profitable. Eventually, the windowing system will break down and consumer choice will emerge, but as of now, it does not appear to be very close to happening.

Sling

A rapidly emerging trend is for consumers to become "cable cutters". The concept is simple: cancel all your cable channels and retain only high-speed Internet service. You can then use Amazon Prime, Apple TV, Hulu, Roku, and other devices or services to watch video content over the Internet. Sling TV, an Englewood, Colorado based live TV service, allows subscribers to watch a collection of cable TV channels on their TVs, computers, or mobile devices. Sling's channels are live TV channels, just like those offered by a cable or satellite TV service. The difference is the channels are delivered over the Internet. See table 4 for a list of devices supported by Sling TV.

Table 4. Devices Supported by Sling TV

Sling Device
AirTV Player
Apple TV
Amazon Fire TV
Amazon Mobile
Android TV
Android
Chromecast
iPhone and iPad
Mac
Windows 10 PC
Roku
Xbox One

Internet connected devices provide access to movies, but a shortcoming is the lack of many cable TV programs. The new Sling solution fills the gap by providing a

Table 5. Sling Orange Channels

Sling Orange Channels
A&E
AMC
AXS TV
BBC America
Bloomberg TV
Cartoon Network/Adult Swim
Cheddar
CNN
Comedy Central
Disney Channel
El Rey Network
ESPN
ESPN 2
Flama
Food Network
Freeform
Galavision
HGTV
History
IFC
Lifetime
Local Now
Maker
Newsy
Polaris
TBS
TNT
Travel Channel
Viceland

less expensive bundle with no multi-year contract. The basic Sling offering, called Sling Orange, includes ESPN, AMC, CNN, Comedy Central, Disney, TNT, food channel, A&E, and a couple dozen more. See table 5 for a list of the Orange lineup as of August 2017. The monthly subscription fee ranges from $20 per month for 30 channels to $40 per month for 50 channels.

For many consumers, Sling TV can be a good replacement for the huge bundles which cable companies offer for $120 or more per month. The cable companies typically have very large channel lineups. For example, where I live, Comcast offers 1,980 channels, almost all of which I never watch.

There are dozens of other sources of movies you can stream without the cable companies including Crackle, Hoopla, Hula, SnagFilms, and Yideo. There is no shortage of video content. The problem is the various restrictions Hollywood and cable companies have built to limit consumer choice. Hopefully, competition from Sling and others will eventually break down the walls and consumers will be able to select what they want, when they want it, what device they want to watch it on, and how much they are willing to pay.

Home Video

It is no longer necessary to drive to a cinema theatre to have an enjoyable movie experience. LED screens with beautiful color have become affordable, and movie watching at home is practical on most any budget. Sixty-five inch screens are no longer as large as they once seemed. With video projectors, it is possible to fill almost any size wall. Built-in speakers are not very good, but surround receivers and streaming systems such as Sonos can bring the audio quality and experience up to the level of the video. Innovation has enabled 3-D TV and curved TV screens, although neither of these have gained significant consumer adoption.

New formats have advanced the image quality of the latest TVs. The previous standard was called 1080P. It is shorthand for high-definition TV (HDTV) video display mode. A 1080P TV has 1,080 rows of pixels with each row containing 1,920 pixels. Most new TVs now use 4K resolution. Also called Ultra High Definition (UHD), a 4K TV model has four pixels for every pixel of a 1080P TV. The result is sharper clarity. The brilliant high-quality video is striking. An enhancement to 4K TVs is high dynamic range (HDR). HDR uses software to enhance the contrast and color profile. HDR, in theory, can show brighter highlights in color images and more detail in dark colors, although some experts are skeptical about whether the benefits can be discerned.[51]

A home attitude can be applied to the video experience in a similar fashion as described for home audio. Suppose you finished dinner and decided to go downstairs to the recreation room and watch a movie. You have a keypad on the wall which also serves as the power switch for the kitchen table. The keypad has four buttons. You press the button labeled Movie. The single button press initiates a series of actions. Music begins to play in the kitchen and recreation rooms. The stairway to the recreation room lights come on. As you go down the stairs, the kitchen table lights go off. When you arrive in the recreation room, the lights and TV are already on. As you sit down, the stairway lights go off and the recreation room lights gradually dim while you are making your movie selection.

When your selection begins to play, the music in the kitchen and recreation room turns off. Halfway through the movie, you decide to take a short break. You take your smartphone out of your pocket and press the pause movie button in your home automation app. The movie pauses, music begins to play in the recreation room and kitchen, the lights in the recreation room come on at 50% brightness, and the stairway lights to the kitchen come on. When you are ready to resume watching the movie, you press the Continue Movie button in the home automation app. The music stops, the lights dim, and the movie resumes. When the movie is over, a tap on the smartphone will bring up the lights and start some music. The home automation system can also make an announcement to your guests through the sound system with a voice and language of your choice. It could say, "The movie is over. Please join us for coffee and dessert upstairs."

The timing between the various actions can be refined based on your experience. Scenes are a fundamental part of home automation. The press of a button in an app or on a wall keypad trigger the scene to begin the steps you have programmed. This is the essence of applying a home attitude to make your home a smart home.

Games

People of all ages enjoy interactive electronic games. These can be as simple as selecting an app on a smartphone, but game playing can be quite sophisticated with powerful computers such as the Xbox, PlayStation, or Apple TV. Multiple players can compete in various life-like scenarios such as adventures in the jungle or war games. Although gaming consoles can be connected to a home automation system and act as a hub, the biggest potential of the consoles is games.

I have never been a gamer because video games make me dizzy. However, I have gotten hooked playing Pokémon GO on the iPhone 7+. Pokémon Go is a free-to-play, location-based augmented reality game. The game involves more than 200 fictional creatures called Pokémon. The player is referred to as a Pokémon Trainer, whose mission is to catch the Pokémon for sport. Location-based means the game uses GPS to display a map of your exact location as you walk around in search of the Pokémon. Augmented reality (AR) is technology which superimposes a computer-generated image on a user's view of the real world, thus providing a composite view. For example, you can display Pokémon from your smartphone to the kitchen table. I like several things about the game.

The technology is impressive. Thinking about millions of people playing the game at the same time with all the results and triggers happening in the cloud is amazing. The graphics and animations are very professional and entertaining. Lastly, Pokémon Go fits nicely with my habit of daily walks. You do have to be careful you don't walk into a tree or a moving car. As of October 15, 2017, I have caught 201 different Pokémon, earned more than four million experience points, dozens of medals, and reached Level 33. Not bad for somebody 3-4 times older than the average Pokémon Go player.

Niantic, Inc. spun out of Google in October 2015 in partnership with Nintendo and The Pokémon Company. The company says its mission is,

> to delight our customers with innovative entertainment experiences which blend works of the imagination with the real world to enable our users to have fun while visiting new places, learning about the world around them, and meeting new friend.[52] The company's systems utilize "high throughput real-time geospatial querying and indexing techniques". The cloud servers process hundreds of millions of game actions per day as people interact with real and virtual objects in the physical world.

Games are serious stuff, especially those built with 3D, augmented reality (AR), and artificial intelligence (AI) technologies. Some of the job openings currently available at Niantic Labs include Production Engineer, Server Infrastructure Software Engineer, Machine Learning Engineer (AI), Mobile Software Engineer,

Unity (a multiplatform game development tool) Game Developer, Quality Assurance Engineer, Operations Manager, and Unity Technical Artist.

The financial side of games is impressive. The video game market's total digital revenue increased 13% year-over-year in November 2016 to $6.7 billion. This was for one month. Although Pokémon GO is free, gamers can make in-app purchases for numerous kinds of "items" such as Poké Balls, which are used to capture Pokémon. In its second-quarter earnings report, Nintendo disclosed it made $115 million in licensing fees from Pokémon GO. As of February 2017, the game has accumulated $1 billion in revenue since it launched.[53]

Virtual Reality

An emerging form of entertainment, although not integrated with a smart home, is virtual reality (VR). VR is a computer-generated simulation of a three-dimensional image or environment enabling a human to interact using specialized electronic equipment such as VR goggles, a helmet with a screen inside, or gloves fitted with sensors. The result is the human has a seemingly real physical experience.

My first experience with VR was in the mid to late 1990s when I visited the Argonne National Laboratory, a non-profit research laboratory operated by the University of Chicago for the Department of Energy (DOE). As part of a tour of the facility, I was fitted with VR goggles and VR gloves. They were archaic by today's standards, with long cables connecting the goggles and gloves to a large computer. I was escorted into a room, and asked what I saw in the far corner. There was a Golden Labrador Retriever sitting on his hind legs. I walked over to the dog and petted him on the head. As I did this, the dog wagged its tail. I was then asked to remove the goggles. There was nothing there. The dog appeared in virtual reality.

VR has been one of the next big things for quite a few years, but still has not established itself as a major player in the world of consumer entertainment. Oculus cofounder Palmer Luckey developed VR goggles in a tech startup for which Facebook paid $2 billion in 2014. The acquisition has not been successful so far, and Luckey has left Facebook.

Google has introduced a technology called Cardboard which allows anyone to build a virtual reality viewer from a few pieces of cardboard. A smartphone fits inside the viewer. Developers create virtual reality content such as a walk through a major city or a tour in a museum. The consumer looks through the viewer and sees a stereoscopic 3D image with a wide field of view. As the viewer is turned from side to side and up and down, the view changes as though you were walking through the scene. If you are on a virtual tour of Paris, and tilt your head back when in view of the Eiffel Tower, you can see the top of it. You must see it to believe it.

Little Star Media, Inc. is a New York startup which has developed a VR platform. Customers can use the platform to create videos which give the consumer the ability to fully explore 360-degree videos from a web browser or mobile device. Using a Little Star player in a browser, websites can embed immersive videos into their site and allow visitors to have a VR experience with no viewer or goggles.

If you have a smartphone, you can make a Cardboard viewer for less than $10, or buy one already assembled for not much more. The lower cost viewers use the smartphone to deliver the images. You insert the smartphone into the viewer and a strap can keep the viewer on your head as you take a tour. More sophisticated viewers are more than $500 and connect by a cable to a PlayStation or PC.

Augmented Reality

Conspicuously absent from the VR space is Apple, which has put its focus on augmented reality (AR). At its September 12 announcement of new iPhones, Apple demonstrated a handful of impressive AR implementations. Its ARKit will enable developers to harness the power of Apple's new chips and cameras. The company sees great potential in many areas including gaming and sports. Experts believe Apple's ARKit will have millions of consumers with more recent iPhones using AR.[54] The technology allows an app to display a high-fidelity visual in a real world setting viewable through the iPhone's camera lens. For example, if you are shopping for a new table lamp, you could see how it would look on your own table. Konrad Gulla, CEO of Keeeb, Inc. and a tech enthusiast is bullish on AR. He said,

> VR with home automation has a limited future, but AR has fantastic potential. In the future, playing an AR game in your smart home with a Microsoft HoloLens-like device, which would look more like normal glasses or even contact lenses, will let the villain of the game sit next to you on the couch. Life-like animals can hide under your furniture. Combine the AR with your multi-color lighting system and the immersive experience will be perfect.

Summary

Consumers have a wide choice of home entertainment ranging from listening to a song, watching a movie, enjoying a game, or participating in a virtual reality tour of places they have never been before. With audio and video home entertainment, the adoption of a home attitude can enable you to automate your entertainment and integrate it with your smart home systems. In the next chapter, I will discuss how the integration of home security with your automation system can enhance your security while you are enjoying the entertainment.

CHAPTER 4
Home Security

D ata from an FBI crime report suggests one in thirty-six homes in the United States will be burglarized this year.[55] The result would be a loss of $2,230 per break-in totaling $4.7 billion in property losses.[56] The financial loss may be surpassed by psychological costs. The victims of burglaries may feel they have been personally violated and live in fear of a future break-in for years. Home security should be a top priority for all homeowners and renters.

Effective home security requires three essential elements: the security hardware in and around the home, diligent homeowner security practices, and intelligent technology to assist in detection and responses. Any one element without the others will not provide the best security. The hardware includes an alarm system, sensors for doors, motion detectors, locks, lighting, and security cameras. The personal security practices include ensuring doors are locked, alarms are activated, windows are closed, and no extra keys are hidden under the welcome mat. Assuming the first two security elements are in place, the rest is up to a security system with a home attitude. A smart home will provide much more than the basics of security.

The Basics

A basic home security system includes sensors for all doors and windows on the first level of the home. If the security system is armed, and a door or window is opened, the intrusion creates a "fault". The fault simply means an electrical contact was broken. A typical door or window contact is shown in figure 2. The contact consists of two cylindrical pieces of plastic, which are inserted into a hole bored into a door or window. One piece of the contact includes a sensor and a wire. The other piece includes a magnet. The technology is simple. When the magnet touches the sensor, a closed circuit is created. When the sensor is separated from the magnet, a signal is sent to the alarm control panel via the wire, signifying a fault has occurred. Immediately after the fault, the security system will use the home landline or cellular alternative to call the security company, which, in turn, will call the homeowner. If there is no answer, or if the person answering cannot confirm a security question, the security company will contact the police.

Door and window sensors come in a variety of designs. The one described above is desirable because it is hidden. A less expensive approach is to use contacts which can be mounted on doors and windows without drilling a hole. Wireless sensors are also available, such as the HomeSeer HS-DS100+ Z-Wave door/window sensor. See figure 3.

Figure 2. Ademco Recessed Contact. By Honeywell International.

Some security companies, who install wired contacts for a living, say the wired systems are more dependable. Wireless sensors have the obvious advantage of not requiring wires to be installed from every door and window to the central alarm panel. Wireless sensors are particularly desirable when remodeling a home or when in a rented apartment or condo. Another advantage of wireless is the ability to include locations where it is difficult to run wires, such as an outside shed. One disadvantage of wireless sensors is they are visible to the homeowner and guests. Another disadvantage is the occasional need to replace batteries in the sensors.

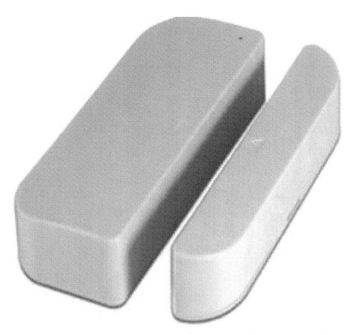

Figure 3. Z-Wave door/window sensor. By HomeSeer.

Enhancements

The basics described provide a rudimentary level of protection against burglary. Unfortunately, the bad guys can be quite sophisticated, and stronger levels of security protection are a good idea. In the sections to follow, I will discuss virtual zones, motion detection, surveillance, and panic buttons. Burglary is not the only protection you should expect from your security system. I will also discuss water leak and carbon dioxide detection. A further set of enhancements can be derived from applying a home attitude. I will discuss how a smart home can help you be as secure as you want to be.

Virtual Zones

The basic building block of a home security system is a zone. Each sensor in the system is assigned to a zone. Although some of the newest security systems are much more flexible than in the past, the typical security system today has a fixed number of zones. For example, Toll Brothers Inc., the Horsham, Pennsylvania luxury home builder, which built more than 6,000 homes in 2016, offers an Ademco

16-zone security system in each home. See table 6 for a list of typical zones a home may have.

Table 6. List of Typical Home Security System Zones

List of Typical Home Security Zones	
1	Front Door
2	Garage Door
3	Back Door
4	Left Living Room Window
5	Right Living Room Window
6	Center Living Room Window
7	Left Dining Room Window
8	Right Dining Room Window
9	Center Dining Room Window
10	Motion Detector - Kitchen
11	Motion Detector - Front Hall
12	Smoke and CO Detector - Upstairs
13	Smoke and CO Detector - Downstairs
14	Police Alert
15	Medical Alert
16	Fire Alert

When a door opens, motion is detected, an alert button is pushed, or a sensor detects something, a zone fault occurs. The fault is displayed on the alarm panel; e.g. Zone 5, meaning someone opened the right living room window. If the security system is armed at the time, the security monitoring company is notified immediately.

While 16 zones are enough for most homes, some homeowners may choose to make the additional investment to get a security system with 32 zones. However, an alternative is to add a home attitude by creating some virtual zones. For example, you might want to add a zone in a place where it may be difficult to get wiring back to the security panel, or when there are no more zones available.

Examples might include a storage shed in the backyard or a garden gate. For these situations, you can easily add wireless window/door sensors. You might also have an afterthought to add a virtual zone near a wine cellar or a gun cabinet, areas you might not want children or visitors to approach even if the security system is not armed.

Any virtual zone can operate the same as the 16 zones in your security panel. By connecting the security panel to your home automation system, every zone, real or virtual, becomes a device in your smart home, and you will have complete flexibility to cause zone faults to trigger an action or action group. For example, if someone approaches the wine cellar, you can have your hub make an announcement through your home audio system. "Someone has just approached the wine cellar".

Motion Detection

The detection of a door or window breach is reliable, but a clever intruder might find another way into your home. Someone could possibly get in through a basement access or even cut a hole through the roof. I remember many years ago watching an episode of a thriller movie where a bad guy attached a large suction cup handle to a sliding glass door. He then used a glass cutter to etch a large circle in the glass. He then lifted the circular piece of glass out, and then stepped through the hole into the home. No door or window was opened, but the intruder gained entry to the home.

Motion detectors rely on the fact any warm-blooded animal emits infrared radiation. An infrared motion sensor contains a thin film of pyroelectric material which detects the infrared radiation of a person and causes a fault in the security system. The detectors can cover a 90-degree angle and a distance up to 30 feet. Careful placement of just a few detectors can cover an entire home. Motion detector faults can initiate any number of different actions. See table 7 for some examples.

A further refinement to detect intrusion is the use of glass break sensors.[57] These sensors are listening devices which use an audio microphone to detect the acoustical frequency which is generated by the sound of breaking glass. Glass break sensors typically have a range of 20 feet in any direction. A single sensor can often cover a sliding glass door plus the windows in a kitchen and family room area. The combination of motion and glass detectors can add an extra layer of protection.

Table 7. Possible Actions Following Detected Intrusion

Possible Actions Following Detected Intrusion
Call the homeowner and designated backups
Call the police if no answer or authentication failure
Activate bright lights to deter an intruder
Sound an alarm
Broadcast a loud message throughout the home
Send text message to homeowner and designated contacts
Send video footage from a surveillance camera

One caution with motion detectors is the use of the Stay mode, a feature of most security systems. The purpose of the Stay mode is to allow you to arm your system while you remain in the home. The feature will deactivate motion detection so you or your pets do not trigger the alarm. The disadvantage is your home is not fully protected while you are at home.

Motion detectors can also be used outside. When a car or a person enters the driveway, or approaches the garage, flood lights can be activated. Multiple motion detectors can activate flood lights around the entire perimeter of a home.

Surveillance

Surveillance means to closely observe activity. In the context of security systems, surveillance is conducted using video cameras. Although surveillance is often used to look for suspicious or intrusive activity, it can also be used to watch children playing in the yard or snow falling on the ground.

Like most consumer electronic devices, surveillance cameras have been declining in price and rising in functionality. Reolink Digital Technology Co, based in Shenzhen, China, started in 2009. The innovative company makes a wide range of security products for consumers and business owners. Reolink's "Plug and Play" technology is of particular value to do-it-yourself users. It eliminates the requirement for reconfigurations or adjustments. Having used many frustrating surveillance cameras over the years, I found this a welcome innovation. The steps to get the camera up and running are so simple, it seems something is missing. Take the

Reolink camera out of the box and plug it into your router. Install the Reolink app on your smartphone. Using the app, scan the QR code (like a bar code) printed on the camera label, and the app sets up the camera. It then gives you the option to add a Wi-Fi connection. That's it. The camera is up and running.

The most recent innovative product from Reolink is the Reolink Argus. The new Reolink is 100% Wire-Free and can be used indoors or outdoors. Advanced power-saving technology allows the camera to operate for up to six months on two batteries. An infrared sensor detects human movements and a built-in microphone and speaker enable you to have a two-way conversation with a person the camera sees. Unlike earlier cameras which had marginal video quality, the new Reolink Argus has 1080P full HD image quality, including night vision. When the camera sees something moving, the motion event can be recorded on a memory card. The Reolink Argus became available in August 2017 with a retail price of $99.99. I have found the Argus to be very reliable. By adjusting the various settings, you can extend battery life significantly.

A further security enhancement is to have multiple cameras all around the home. Smarthome.com offers a comprehensive security package with eight cameras. Each camera captures video in Full HD 1080p resolution. Night vision camera technology provides around-the-clock protection and peace of mind. The system includes a digital video recorder so you can view what is going on from each camera from your desktop, laptop, tablet, or smartphone. The weatherproof cameras, in effect, provide eight extra sets of eyes to monitor your property. If a camera sees something move, it takes snapshots, and emails them to you. The system can store up to 30 days of video, depending on the picture size and other parameters you choose.

As the technology for surveillance gets more advanced and the cost continues to come down, it will empower consumers to know exactly who is coming and going around the entire perimeter of their home. The technology is powerful, but I recommend considering the pros and cons of deploying multiple surveillance cameras before making the investment.

The most important advantage would be the deterrence of crime.[58] A thief seeing prominent surveillance cameras, and realizing every movement will be recorded, may think twice before committing a crime. Hidden cameras can be used to increase the odds of catching thieves in the act. Hidden cameras or not, if a crime is committed, the recorded video will provide the evidence. The cameras can also provide recorded audio of criminals planning their approach.

A second advantage of surveillance is for non-criminal activities. For example, the cameras can confirm the arrival of friends, family, or service personnel. After seeing a repairman has arrived, the consumer can unlock the door remotely to allow the repairman access to the home.

The primary disadvantages of surveillance cameras all relate to privacy. Friends and family may be uncomfortable knowing their comings and goings are being recorded. A second disadvantage is the cost. Although the cost of cameras has declined, the professional installation and maintenance costs have increased with inflation. It would be easy to spend thousands of dollars on a high-quality system installation. There can be vulnerabilities with the use of surveillance cameras. If they are connected to the Internet for remote viewing, and if strong password security is not implemented, a camera or multi-camera control unit could be hacked and turned off.

In summary, the use of sophisticated video surveillance cameras can enhance security, but the pros and cons should be considered. The most important security is to have the perimeter of your smart home secured with sensors at all entry points and to have the interior fully covered with motion sensors. Deployment of surveillance cameras needs to consider the additional societal and privacy factors.

Alert Buttons

In 1989, LifeCall, a medical alert services company began running television commercials containing a scene with an elderly woman. The woman, Mrs. Fletcher, took a fall in the bathroom, and said, "I've fallen, and I can't get up!" Fortunately, Mrs. Fletcher was wearing a LifeCall medical alert pendant. The dispatcher called in response to her pushing a button on the pendant and told her he was sending help. The phrase, "I've fallen, and I can't get up!" is currently a registered trademark of Life Alert Emergency Response, Inc.[59]

Home security alarm keypads typically have alert buttons labeled Police, Fire, Medical. If you push the Police button, the alarm monitoring company receives a signal and immediately dispatches the local police department to your home. The Fire button will result in dispatch of the nearest fire station, and the Medical button will result in dispatch of an ambulance. The keypad can be useful for an emergency but, in many cases, such as a fall, a homeowner may not be able to get to the alarm keypad.

Numerous companies offer devices to enable a consumer in the home to create an alert and request assistance from a variety of devices such as pendants, wrist bands, watches and, of course, smartphones. After pushing a button, the device connects with a call center such as LifeCall or Life Alert which are staffed with trained specialists who know how to deal with emergencies. Some devices allow people to answer incoming calls from a dispatcher when the telephone is out of reach.

An alternate way to contact emergency services is to text 911. The FCC Text-to-911 program can enable a homeowner to send a text message to reach 911 emergency call takers from a mobile phone, laptop, or desktop computer.[60] A person without a mobile device can send a text message using email. The major wireless service providers support texting 911 in areas where dispatchers are equipped to receive messages. For example, using AT&T, a person could send an email to 911@txt.att.net.[1] Unfortunately, most parts of the U.S. don't yet offer the Text-to-911 option, so you should make a voice call to contact 911 during an emergency whenever possible. See table 8 for how to send a 911 text message by email.

Table 8. How to Send a 911 Text Message by Email

How to Send a 911 Text Message by Email	
AT&T	911@txt.att.net.
Sprint	911@messaging.sprintpcs.com
T-Mobile	911@tmomail.net)
Verizon	911@vtext.com
Virgin Mobile	911@vmobl.com.

The various devices and services described are useful for emergencies, but a button push could do much more when the process is infused with a home attitude. Instead of a single purpose button, a smart home button can add flexibility and customization. Home automation "buttons" come in many forms: single button, a mini remote control with multiple buttons, a button on a pendant, or buttons which can be stuck on a shower wall or kitchen cabinet. While

---

[1] You can send a text message to anyone if you know which provider they use. Simply substitute the person's mobile number for 911 in the addresses shown in table 8.

the medical alert services typically offer a specific button tied to their service, a home automation solution is much more generic and can be tailored to exactly what you want.

When a designated emergency button is pressed on any device, the home automation hub receives the signal and executes an action group. The action group can include as many actions as you want. See table 9 for an example of some possibilities.

Table 9. Example of an Emergency Action Group

Example of an Emergency Action Group
Send a text or email to 911
Send a text message to a family member
Send an email to a different family member
Turn all the lights in the house on
Activate the police, fire, or medical button on security alarm panel
Cause all exterior lights to blink on and off
Activate burglary alarm

Home security has many dimensions. The important thing is to cover the basics: door and window sensors, motion detectors, an alert button, and a monitoring service. **As** As discussed earlier, you can add numerous enhancements. Like insurance, you can get minimum protection or coverage for every possibility. It is a personal choice.

Leak Detection

We have had a family vacation cottage on a lake in the mountains of Pennsylvania for four decades. In 2012, we arrived just before Memorial Day weekend. We opened the front door and saw water everywhere. The tubing which connects the ice machine in the refrigerator had come loose and water was pouring onto the floor. The entire cottage was flooded. As we walked down the steps to the lower level to turn off the source of the water, we could see water dripping from the ceiling and running down the walls. The water going down the steps flowed into the guest bedroom and down the hall to other rooms. It was a disaster which could have served as a TV commercial for SERVPRO, a leader in the cleanup and restoration industry.

We hired a similar company, and they brought in fans, pumps, vacuum cleaners, and specialized equipment for mold remediation. Their contract took four pages to describe the scope of the project and what they would be doing. After the remediation, we hired a contractor to remove the carpets, the underlayment flooring, and wall panels. Built-in cabinets had to be removed and later reinstalled after new floors were in place. In summary, it was a disaster and the total cost was approximately $25,000. Fortunately, most of it was covered by insurance, but the disruption to summer plans was significant.

Right after the disaster, I applied some home attitude to the cottage. First, I purchased three Insteon Water Leak Sensors. See figure 4. I placed one sensor on the floor behind the refrigerator, a second at the foot of the hot water tank in the garage, and the third in the laundry room near the pipe which brings water into the cottage.

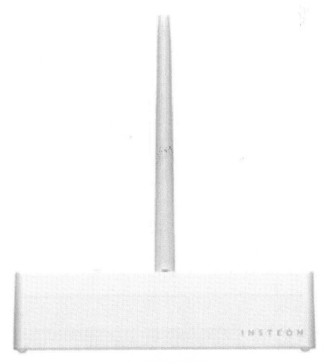

Figure 4. Insteon Water Leak Sensor. By Smarthome.com.

The leak sensors, which include a 10-year battery, have two small metal spherical feet on the bottom. If water touches both feet, an electrical contact is made and the sensor sends a signal to the home automation hub. When the hub receives a signal from any of the water leak sensors, it immediately closes an electrical valve which shuts off water to the home. The hub then sends an email to me, my wife, and our neighbor across the street with a subject line: "Water Leak Detected at Cottage. Action Required." A small investment of $30 to $35 per sensor can save thousands of dollars.

Carbon Dioxide

A teenage boy died, and 14 other people were found unconscious around an indoor pool at a southern Michigan hotel in April 2017.[61] The cause of death was carbon monoxide poisoning. Fire officials said they believed the tragedy resulted from a defective exhaust ventilation system on the pool heater.[62]

Four hundred people per year die from carbon monoxide poisoning, and 20,000 are admitted to the emergency room.[63] Carbon monoxide is often called the silent killer because it has no odor, taste, or color. If you become unconscious while breathing carbon monoxide, the odds of becoming extremely ill or even dying are alarmingly high. Carbon monoxide is the most common type of fatal poisoning in the world,[64] and it is worth knowing more about.

Let's start with oxide. Oxide is the name of a simple compound which includes oxygen plus something else. For example, if you combine an oxygen atom with two hydrogen atoms, you get hydrogen dioxide (H_2O), or water. If you include a nitrogen atom with two oxygen atoms, you get NO_2, or nitrous oxide, commonly called laughing gas. The first part of the name monoxide means one oxygen atom, and the first part of dioxide means two oxygen atoms.

CO is carbon monoxide, and CO_2 is carbon dioxide. The names are similar and they often get confused in articles. They are actually quite different. CO_2 occurs naturally in the atmosphere, and is essential for plant life. It is produced as a byproduct of animal and human respiration. CO_2 is generated by gasoline engines which use a catalytic converter. CO_2 poisoning is rare.[65]

CO does not occur naturally in the atmosphere. It is created when the ventilation system of a fuel-burning appliance is defective and the appliance is starved for oxygen. It can be associated with fireplaces, wood stoves, space heaters, furnaces, or water heaters. CO is generated by any gasoline engine which does not use a catalytic converter.[66]

When you burn dead wood, carbon monoxide is released into the air. When you are outdoors, the CO can easily dissipate. However, when you are indoors, or in a garage or an RV, the CO can reach a toxic level and create a health hazard. The Occupational Safety and Health Administration (OSHA) requires long-term workplace exposure levels of CO to be limited to 50 molecules per million molecules of air. Mild CO poisoning symptoms, such as dizziness or headaches, can occur at 100 parts per million (ppm), and at a concentration of 700 ppm, CO can be life-threatening.[67] The concentration at the indoor pool in southern Michigan was reported to be 800 ppm.[68]

Any home with a gas stove, space heater, or fireplaces should be protected with carbon monoxide detectors on all levels of the home. CO detectors are available with a wide range of features. Detectors have typically required wiring for power, but they are also available with Wi-Fi. Some work with batteries and some use batteries as a backup in case of electrical power loss. Most CO detectors also detect smoke, and some can detect a propane or natural gas leak.

CO detectors are typically monitored by security monitoring services. The timing of the reporting depends on the concentration and duration of the detected CO. For example, a concentration of 50 PPM will trigger an alarm in eight hours. At a concentration of 150 PPM, the alarm will be triggered in 10 to 50 minutes, and at 400 PPM in 4 to 15 minutes.[69]

Smoke, fire, and CO detectors can also be connected to a home automation system. If an alarm is triggered, a message will be received by the hub, which can then trigger an action group consisting of various notifications such as with the leak detector example. The smoke, fire, and CO detectors are not protecting against unwanted intrusions of people, but of situations hazardous to home occupant health and safety.

Summary

Homeowners have access to a wide range of devices which can enhance their safety and security. Inexpensive sensors can detect entry, motion, the sound of breaking glass, water leaks, smoke, fire, and carbon monoxide. Sophisticated cameras can provide surveillance of the driveway or the entire perimeter of the home.

Alert buttons can dispatch police, fire, or medical personnel in case of emergency. These capabilities can be managed as part of a standard home security system and monitoring service, or by applying a home attitude, they can be integrated with a home automation system. With the smart home approach, many enhancements are possible which can enable you to customize your safety and security to whatever level you may want. In a later chapter, I will discuss network security, which affects who can access your smart home. In the next chapter, I will discuss how a home attitude can be applied to your home's energy usage.

CHAPTER 5
Energy Efficiency

A couple of years ago, a friend of mine told me his electric utility company had sent him an email saying the electrical consumption in a vacant home he had for sale was out of line compared to other homes in the neighborhood. I am not sure what algorithm they use to make this determination, but it sounded convincing. He decided to wait for the following month's update. It showed the same thing, a significant variation compared to other homes in the neighborhood.

My friend retrieved electric bills from the prior year when the home was occupied, and found the current electrical consumption was significantly higher than the prior year. In financial terms, the current electric bills were running a few hundred dollars per month more than when the home was occupied. Something in the home must be running. We went to the home and checked every room, the basement, garage, and attic. Some appliances and electronic items can be using electricity even when they are turned off. We unplugged them all. We checked the thermostats to make sure the home was not unnecessarily being heated or cooled. He then waited for the following month's bill.

Once again, the kilowatt hours of electrical power consumed was significantly higher than for the same month of the prior year. Something had to be running my friend was not aware of. I called another friend who is quite knowledgeable about all aspects of utilities for homes. His first question was about where is the source of water for the home. The water source was a well on the property. He suggested calling a well specialist. An inspection by the well specialist revealed the well pump was clogged with iron rust and unable to deliver the required water pressure. As a result, the well control kept the well pump on continuously. The new pump and related labor was expensive, but not nearly as much as continuing to run the pump 24x7 indefinitely.

Americans have had a major dependency on electricity in the home for more than 75 years. When electrical power to the home is lost, we can face major inconveniences. When we are using electricity unnecessarily, we may not notice it until we get the bill. In this chapter, I will discuss how a smart home can give us feedback on our energy usage.

Monitoring Energy Usage

The American Housing Survey (AHS) is sponsored by the Department of Housing and Urban Development (HUD) and conducted by the U.S. Census Bureau. The survey is the most comprehensive national housing survey in the United States. I extracted a subset of the data, excluding the very high and very low incomes. The data show housing cost is approximately $1,400 per month or 36% of income.[70] Nearly $300, more than 20% of income, is spent on home energy usage, split roughly equally on electricity and oil or gas. An expenditure of nearly $4,000 per year, and much more for many, suggests monitoring the usage of home energy can be quite important and easily justified.

Electricity

When I was at IBM, some of the product laboratories would loan me new equipment they had developed. I had been the Vice President of Marketing for the IBM PC Company when we launched the IBM ThinkPad in 1994, and I evangelized about the product in many speeches and blog postings. The other labs knew I would do the same with their products. Some of the products loaned to me were better suited for small businesses, rather than for home use, but I was happy to have them and use them.

One of the products was an IBM equipment rack about as big as a medium-sized refrigerator. It fit nicely in my basement. Inside of the rack, I had a server and a large array of computer storage, which I loaded up with music. In effect, I had an early version of my own cloud computing infrastructure and streaming music service. The installation was overkill for a consumer, but fun to learn about, write about, talk about, and use. I had it all tied in with the home automation system. I could select music tracks from keypads around the home. This was before Amazon Web Services, the iPhone and iTunes, and other music services.

One day I was showing the equipment rack to a friend. He put his hand on the top and noted the rack was quite warm. This was not surprising, given the industrial scale equipment inside. Then it dawned on me, I wonder how much electricity I am using. I looked around for ways to monitor this, and discovered P3 International, a New York based company, which had developed a product called the Kill A Watt® EZ.[71] The innovative product has a user-friendly power meter which enables people to calculate the cost to use their home appliances. See figure 5.

Figure 5. Kill A Watt® EZ. By P3 International.

The company says, "Now you can cut your energy costs and find out what appliances are actually worth keeping plugged in." I was intrigued. The Kill A Watt® EZ is incredibly easy to use. You simply plug it into a receptacle, and then plug the appliance you want to evaluate into the Kill O Watt. You enter the cost per kilowatt you pay to your electricity provider. The large LCD display will measure the consumption by the kilowatt-hour, the same as your local utility, and then show the operating cost of your appliance or, in my case, an equipment rack full of IBM equipment. The display calculates the immediate cost and forecasts what it will be by the week, month, and year.

When I plugged the IBM equipment rack into the Kill A Watt® EZ, the numbers stunned me. The projected cost of operating the rack exceeded $100 per month. Today, you can buy a lot of streaming music and cloud services for much less than half of that. Since the equipment was all fully depreciated and of no further interest to IBM, I sold the equipment on eBay for a modest amount. Reducing the amount of energy can not only save money, but also save the environment. I donated the eBay proceeds to Carbonfund.org to help offset my carbon footprint.[72]

The Kill A Watt® EZ only measures the energy used by devices plugged directly into the meter. There is also good reason to monitor the total amount of energy used for everything in the home. This could enable you to see the collective effect of using home automation to turn unnecessary lights off and removing or replacing inefficient appliances.

Aeon Labs, was founded in 2006 and is based in Silicon Valley, California. The company designs, develops, and manufactures products for home automation. The company describes its technology as, "Designed to improve the pleasure we derive from the home and office spaces in which we spend most of our time, and to save energy."[73]

An example of the latter goal is the Aeotec Home Energy Meter. When connected to your home electrical panel, the device will monitor the total consumption of electricity used by your home. See figure 6 for a picture of the Home Energy Meter. The device reports your energy use to your home automation hub real-time, so you can see how much total electricity you use and when you use it.

The Home Energy Meter must be wired into your electrical panel. This should only be done by a professional electrician to ensure a safe installation. The electrical usage data is transmitted to your hub up to 300 feet away via wireless communication. Once you have the data, there is a lot you can do with it. You may want

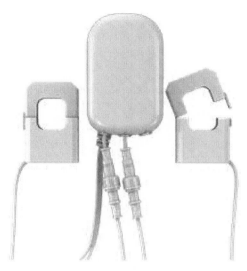

Figure 6. Aeotec Home Energy Meter. By Aeon Labs.

to see a daily, weekly, or monthly email showing the usage. You can compare your monthly usage with what your electricity provider charges you. Most importantly, you can see the effect of your efforts to use energy efficiently and save as much money as possible.

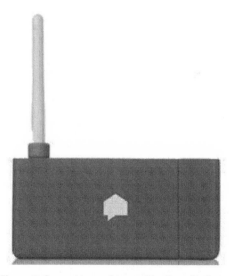

Figure 7. Sense Home Energy Monitor. By Sense

Cambridge, MA based Sense has developed a home electricity monitoring system which can identify how much energy each individual home device is consuming. The Sense Home Energy Monitor can be installed into your electrical panel like the Aeotec product. It monitors electricity consumed by home appliances, and can identify them uniquely based on algorithms it has developed.[74] See figure 7. The company provides a specially designed mobile application to allow consumers to see appliances which may be defective or using electricity excessively. At a cost of $299 for the device plus professional electrician installation cost, the Sense monitor may take some time to pay for itself through saved electricity.

Backup Generator

The family vacation cottage in the mountains of Pennsylvania endures a lot of power outages. It is in a remote area with above ground power lines which are vulnerable to wind and storms. In 2007, I decided to invest in a propane powered backup generator. Founded in 1959, Generac was the first to develop affordable home standby generators, and they claim to be the #1 manufacturer. I found a local contractor who was able to install the generator on a concrete slab under an elevated storage shed in my side yard.

The generator I selected only covers critical areas of the home: refrigerator, heating system, the most important lighting and outlets, security system, cable modem and router, and the home automation system. One nice feature of the generator is an automatic weekly test cycle. On the day and time you select, the generator runs for 15 minutes and then shuts down. The weekly exercise of the engine prevents its oil seal from drying out and damaging the generator. It also provides some charging of the starter battery.

The weekly engine exercise works well, assuming the battery can start the engine. If I am not at the cottage, I have no way of knowing if the weekly exercise occurred. I have had several occasions over the past five years where the exercise did not occur because the battery was dead. A power failure occurred, and the generator did not do its job. I got a freeze warning message from my home automation system and, fortunately, my neighbor was at home. He checked the home and found the power had been restored. A contractor came later and replaced the worn-out battery.

Although I have avoided catastrophes, I concluded there must be a way to monitor the generator and be sure it was able to make its weekly run. Generac offers a $279 Mobile Link™ Remote Monitoring feature which allows you to monitor the status of your generator from anywhere using a smartphone, tablet, or desktop app. It is a well-designed feature, but it is expensive and it is not available for any generators older than 2008 (mine is 2007). Replacing the generator would be very difficult and expensive.

There must be a way some home attitude could be applied to the problem. I tried several experiments with the home automation system and developed a solution. I purchased an inexpensive wireless vibration sensor and used Velcro® to attach it to the top lid of the generator. When the generator runs on Saturday morning, the sensor detects the vibration and sends a wireless signal to my home automation hub in the smart home. The signal causes a trigger which, in turn, starts an action which sends me an email saying, "The generator just started". I created a variable in the hub called Generator Ran and set the value to "No". When the generator runs, the hub sets the variable to "Yes". An hour after the generator was supposed to run, the hub checks the variable. If it is still "No", I get an email saying, "The generator did not run". I can then take appropriate action.

Fuel Oil

The level of oil in a fuel oil tank has historically been measured with wooden or metal dipsticks, tank clocks, dial gauges, or floats inside the oil tank. Innovation is long overdue, but Stutensee, Germany based Inno-Tec GmbH has filled the gap with an innovative product called the Proteus EcoMeter.[75] The EcoMeter uses ultrasonic technology to measure the level of fuel in an oil tank. The company says the ultrasonic technology provides measurements up to 10 times more accurately than comparable mechanical or electronic measuring devices. Fuel oil suppliers can monitor the filling level in the tank accurately without having to open the tank, thereby avoiding spills.

Ultrasound refers to sound waves with frequencies much higher than what humans can hear. The concept behind ultrasound is quite simple. The sound waves are emitted and they travel until they hit something and bounce back. By measuring the time of the round trip, it is possible to measure the distance. It is used in many applications. The Tesla Model S was the first car to use ultrasound for long range sensing. Elon Musk, South African-born Canadian-American business magnate, investor, engineer, and inventor, said the Tesla ultrasound system is, "long-range, offers 360-degree coverage, and establishes a protective cocoon around the car. It can see anything: a small child, a dog. And, it can operate at any speed."[76]

The Proteus EcoMeter consists of a long life, battery-operated, ultrasonic sensor which can detect the level of fuel in a tank. It has a digital radio receiver which is equipped with an easy-to-read LCD display. The transmitter can send the tank level nearly 500 feet to the display. See figure 8 for a picture of the LCD display. The EcoMeter is nicely designed for the consumer. There are no mechanical parts needed to be installed inside the tank and there is no wiring required. The company has sold more than one million EcoMeters in Europe. The cost is approximately $150.

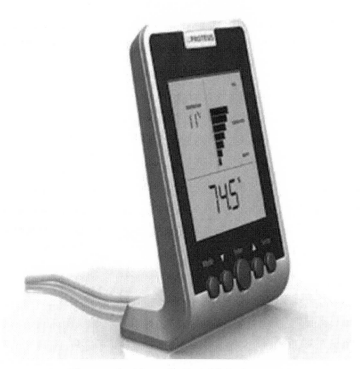

Figure 8. EcoMeter for Fuel Oil. By Proteus.

The EcoMeter can enable you to conveniently monitor your fuel oil level anywhere in the home, and it can be tied in to a home automation system. Adding the home attitude factor enables you to have schedules to check the fuel level at predetermined times. Actions can send you text messages when the fuel is low or when a delivery is made. In the winter, you can combine the fuel oil level with your morning home audio Good Morning greeting.

The EcoMeter monitoring provides more than simply the fuel level. It shows the remaining amount and percentage of fuel in a graphical display, plus it displays the average consumption per day, week, month, and year. The system provides cold weather warnings, a battery status alarm, and a heating cost overview. An additional feature is the monitoring of CO_2 emissions.

Propane

Propane and natural gas expenditures, on average, are 36% less than for fuel oil.[77] Nevertheless, it could be useful to monitor the usage to see the effect of changing thermostats or other energy saving strategies, rather than wait for the bill. Natural gas supply is theoretically infinite, but with a propane tank, it would be good to know when it is running low and when a delivery has been made.

Monitoring of natural gas and propane gas can be done technically, but there are barriers to putting the necessary equipment in place. Companies which make monitoring equipment focus on businesses who have multiple locations and tanks. Their equipment and services are not affordable or justifiable for most consumers. Connecting sensors to tanks or gauges requires some form of electrical connection, and using the words gas and electricity in the same sentence raises significant legal liability issues. In the case of natural gas, it is usually provided by a city or township, and they specifically prohibit the attachment of anything to their gas lines, even inside your home.

Eventually, propane tanks and natural gas meters will probably include a Wi-Fi connected sensor which provides continuous monitoring. However, because of the legal liabilities and utility company restrictions, there are few if any licensed electricians or plumbers looking for monitoring installation business. Until monitoring becomes a standard feature for propane and natural gas, the only way to install a home attitude for energy monitoring is the do-it-yourself approach.

Joe Thomas, a senior software developer at Bloomberg in New York, had been looking for a solution to remotely monitor the level of propane gas in outside tanks at a family vacation home in upstate New York for years. In 2015, he gave up waiting, and decided to create a monitoring solution himself. He developed an approach which has worked reliably for more than two years. Joe can see the percentage full of his propane tank on his smartphone. He also gets text alerts when the level passes thresholds he chooses, such as below 25% or 10%. He gets a text message when the propane company refills the tank so he can prepare to pay the impending bill. More importantly, he gains the comfort to know he can go to his vacation home in the winter and not worry about encountering an empty propane tank.

Joe's solution is not for the faint-hearted. It involves installation of two special hardware items and some expertise with a soldering gun and test equipment. The electronic parts cost approximately $150. Joe is quick to note he is not an electrician or plumber, and he cautions do-it-yourselfers to implement at their own risk. See http://www.rochestergauges.com/pages/locations.html for technical details on the equipment Joe used to develop his elegant solution to detect the level of gas in the propane tank.

At my home, I have two propane tanks buried underground in the backyard. The tanks provide energy for the outdoor grill, kitchen stove, ovens, clothes dryer, hot water, fireplaces, radiant heat in the floors, and a 25,000-watt backup generator. Being so dependent on propane is the reason for two tanks. I had the Viessmann furnace inspected after 15 years, and the service technician told me it was, "Clean as a whistle". Although, over those years, propane may have been slightly more expensive than fuel oil, the total cost, including servicing, has likely been less.

The propane provider, the home automation contractor, and I designed some home attitude into the propane installation. The tanks share a monitoring sensor like what Joe Thomas used. It was installed professionally. The sensor is connected to a small control panel in my basement via an underground cable. The panel is wired into the home automation hub. I can monitor the level of propane on home keypads or remotely via iPhone and iPad.

I can get alerts if the level runs low, but I don't have to worry because the propane provider is monitoring the tanks also. Between two and three AM every morning, a small telephone panel, which is connected to the control panel, places a call to the provider and reports the propane level. When the level gets low, the provider dispatches a truck to deliver propane. This provides benefits to both the propane provider and to me. I like it because it minimizes the number of times the truck ties up my driveway, and the driver must drag a substantial hose through a garden gate and across the lawn. I also get fewer bills to pay than otherwise. The provider likes it because fewer propane deliveries adds efficiency to his operation.

Natural Gas

Because of the restrictions imposed by city or township, which specifically prohibit the attachment of anything to their gas lines, natural gas monitoring is more challenging than propane monitoring. Poul-Henning Kamp, a Danish computer software developer, decided to take on the challenge. He lives in Slagelse, Denmark and is known for his work on various global software initiatives. He has contributed to many open source projects which have improved the quality and security of the Internet. Poul-Henning is a highly accomplished technical expert in many areas. Unfortunately, at this stage, it takes such talent to figure out how to monitor natural gas usage in the home.

Paul-Henning discovered the gauge on his natural gas line incorporated some technology inside which could be used to monitor the usage, but the utility forbids anyone from tinkering with it. Upon closer examination, he noticed the rightmost zero on the gauge was metallic and colored, and he got the idea to find an optical sensor which could recognize the zero. He then designed some electronic circuitry to pick up the signal from the sensor and transfer it to his computer using a modified printer cable. Poul-Henning implemented his innovation in 1996 and has used it to gather a lot of data.[78]

I asked Poul-Henning if he was optimistic about some mainstream monitoring technologies to emerge soon. He said, "Not really. Something new will have to be devised."[79] He sees a solution like what Joe Thomas devised as a possibility for monitoring in the future.

Although Poul-Henning is not optimistic about the near-term availability of inexpensive, easy to use, accurate, and meaningful monitoring technology, he does see monitoring as potentially quite helpful. "If you want to reduce natural gas usage, you have to monitor it." He said, "Natural experiments all over the world have shown up to 10-15% reduced usage, if meters are visible as part of your daily routine, as opposed to being hidden away somewhere." [80] Getting a continuous stream of data from monitoring is theoretically useful, but in practice the data can be hard to interpret correctly.

Poul-Henning believes monitoring data can be useful as a corollary to weather data. For example, by combining the data with wind direction data, it can help plan for where more insulation of your home may be effective in reducing heating needs. See figure 9 for an image of a two-year time-exposure taken by measuring how much sunlight hits various parts of his roof solar panels. It shows which trees shade the panels, and thereby limit the efficiency of his solar heat.

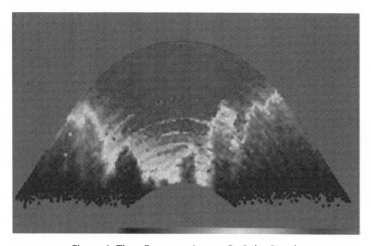

Figure 9. Time Exposure Image. By Solar Panels

Poul-Henning Kamp is a true believer in monitoring everything he can. He has thermocouples to measure the temperature under his home foundation, the heating unit, ventilation system, and at his solar electrical panel. He also has electrical meters on all appliances, and sensors on water, electricity, heating, solar panels, and other things. He enjoys being creative with monitoring and then enjoying the study of his data. He has clearly demonstrated a home attitude, and has set the bar quite high for the rest of us.

Water

The koi pond in our back yard is in open space. There are no trees in the area tall enough to prevent the sun from beating down on the pond and causing algae to thrive, it also causes water to evaporate. The pond has a small waterfall powered by a submersible pump. If the water level gets too low, the pump will gasp and eventually burn out.

There are numerous types of water level sensors which could enable me to monitor the water level, but I decided to take a different approach. I had a plumber but a PVC water line from our basement, under the ground, and into the pond. He put a solenoid valve in the water line in the basement, and I connected the wire cable from the valve to the home automation hub. Rather than automatically sensing the water level of the pond and having the hub turn on the water, we setup a daily schedule. For example, every day at 3 PM, the water would turn on for 10 minutes. We got to know the right time and amount from experience, and the system worked just fine.

However, we started to notice there was not enough water in the pond. We upped the time limit to add more water, but the level continued to be too low. We summoned the pond contractor, and he discovered there was a leak in the rubber mat which lines the bottom of the pond. The water was running into the ground. As it turned out, I am glad we did not have a water level sensor and automatic filling. We may not have noticed the problem and wasted an excessive amount of water. The excess may also have worn out the well pump, while wasting a lot of electricity in the meantime.

Saving water can be good for your budget and a good thing for the environment. Lake Wylie, South Carolina based FortrezZ, has a product called the Flow meter.[81] It connects to your water intake plumbing and sends data to your home automation hub to help you manage water consumption and detect water leaks in your home. Flow Meter tracks water usage and reports gallons used. You can track household water usage and see where your peak consumption is. Flow meter checks the temperature nearby, for example in your basement, to detect freezing conditions and sends temperature alerts. Flow meter tells you if water is flowing or if it's off. It knows if the flow is small, like a running toilet, or if it's large, like a garden hose left on. You can create triggers in your home automation hub to provide alerts. For example, you can create an action in your hub to automatically turn off water if too much water is flowing, or automatically turn water off if it has been running for too long. You can also set limits to turn water off when a set number of gallons have been used for the day.

Summary

In this chapter, I discussed several aspects of home utilities, including electricity, propane, natural gas, fuel oil, water, and backup power generation. I also described techniques for monitoring these critical resources. Not only can monitoring give you piece of mind from knowing what is going on in your home, it can also help you save as much money as possible as a result. Something else to monitor is the weather.

CHAPTER 6
Weather Monitoring

Most of us follow news, sports, politics, financial markets, or weather. Walt Hickey, chief culture writer at FiveThirtyEight, did a survey and found 80% of respondents check a weather report daily. In some geographies, weather conditions can change dramatically by the hour. [82] For example, in Phoenix, Arizona, temperatures often range by as much as 30 degrees during the day. In some areas, it can be dark and rainy part of the day, and clear blue sky for another part of the day.

Hickey found The Weather Channel on TV has dropped to 15% usage as a source of weather information. [83] The default weather app on smartphones or websites have gone from zero 20 years ago to more than 40% usage in 2015. [84] An emerging source of checking on the weather is the smart home.

Weather Underground

Founded in 1995, Weather Underground is based in San Francisco, California as an offshoot of the University of Michigan's Internet weather database. Jeff Masters, a doctoral candidate in meteorology at the university, developed a computer program in 1991 which displayed real-time weather information on the Internet. A few years later Weather Underground developed a web based version which, for the first time, gave millions of Internet users the ability to see a colorful and graphical view of weather conditions and forecasts. [85]

Weather Underground has been innovating access to weather for more than 25 years. In addition to sharing weather information with the world via web pages, smartphones, and tablets, they use weather observations which come from nearly 200,000 personal weather stations worldwide.[86] Weather data travels in both directions, to users and from users. For those with home attitude, Weather Underground has a software program which creates a stream of data which a hobbyist can feed directly into a home automation hub.

Home Weather Stations

Another way to get weather data is to install your own weather station equipment. Qubino, a brand of GOAP, a Slovenian global innovator in home automation, builds cabin automation for the world's biggest cruise ships, and home automation devices which measure indoor and outdoor temperature and humidity, rainfall, wind direction and speed, and wind chill.[87] Qubino devices, which mount on and around your home, measure data and send it to your home automation hub so you can use it to make your home smarter. See figure 10 for a picture of the devices.

With the data at hand, you can incorporate the weather into various triggers and actions. If severe weather is threatening, you can receive a text warning or audio announcement and have the garage door and powered shutters automatically close. As storm conditions change, you can receive update announcements through the home audio system. If the driveway or sidewalks accumulate freezing rain, you can automatically activate underground heating coils and turn on a greenhouse heater.

You can have the day's weather forecast read to you through your home audio system when your Good Morning action takes place. The home automation hub can organize the exact weather data you care about, and place it in a customized format on your smartphone and tablet.

The weather can have a large impact on a home irrigation system. By using a rain sensor plus electronic controls on each irrigation head, you can gain better control. With the irrigation schedule in the home automation hub, you can ensure no irrigation will take place when it is raining. You can supplement your irrigation control with soil resistance sensors. The sensors can determine how much moisture is in the soil. You can use this information to fine tune the watering schedule and avoid watering too little or too much.

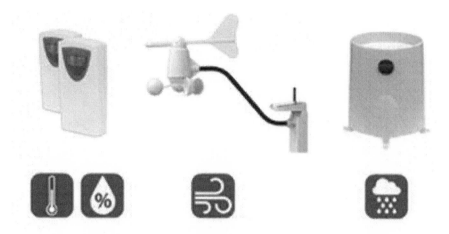

Figure 10. Qubino Weather Monitoring Equipment. By GOAP.

There is nothing you can do to change the weather outside but, inside your home, there is a lot you can do. If the skies blacken during the day, your first-floor lights can turn on automatically, and then turn back off when the brighter skies return. If the sun is beating down on your windows, your solar shades can automatically go down. Although, your thermostats regulate the temperature in a room, sometimes a ceiling fan can be a nice supplement and allow the temperature to be a bit higher with no more discomfort. If the temperature in my home office exceeds 75 degrees, the ceiling fan goes on at medium speed. If the temperature continues to climb after a half-hour, the fan moves up to high speed. When the temperature drops to below 74 degrees, the fan turns off.

Summary

You can think of weather as a phenomenon which is outside of the home. If you apply some home attitude, you can think of weather as something you can monitor, use as input for planning and alerts, and control its effects inside your home. While gaining a sense of awareness and control, you can also enhance your safety and reduce your energy budget. In the next chapter, I will discuss how a home attitude can provide convenience, not safety or savings, just making things a bit easier.

CHAPTER 7
Convenience

P arrot, is a Paris based company which creates, develops, and markets high tech wireless products for the retail and professional markets. The company focuses on three market segments: Civil Drones, Automotive communication and infotainment systems, and Connected Objects. The latter is another name for the Internet of Things (IoT). The Connected Objects Parrot has developed are related to gardening. See figure 11 for a picture of the Parrot Pot.

Figure 11. The Parrot Pot. By Parrot.

The Parrot smart pot will water your plant for up to one month with no human intervention. The automated pot can control watering from your smartphone. For those with home attitude, you can connect the pot to your home automation hub and provide triggers to not water when it is raining, and send alerts to warn you if there should be threatening high winds or storms which could damage your plant.

The Parrot Pot is approximately 8 by 12 inches, and it is super smart. The pot includes sensors to measure soil moisture, fertilizer level, sunlight, and air temperature. The company describes the irrigation system of the smart pot as a "Perfect Drop". Your plant only gets watered when it needs it, and it automatically adapts to the plant's natural life-cycle and adjusts water consumption accordingly. The result gives your plant just the right amount of water at the right time. Parrot says, "Enjoy lush, thriving plants!"

Some readers may find the smart flower pot an over-the-top unnecessary extravagance, or even a useless gadget. Others may find it as innovative, useful, and even a necessity. Some prefer to visit the terrace or deck with a watering can and personally attend to their plants. Others may be frequent travelers who welcome a way to protect and nurture their plants while they are out of town. Both views are fine. A similar dichotomy of views exists for home automation. Some say, why bother? Others say they cannot live without it. I do not argue a home attitude mandates every application of home automation is needed. Some things, like security and heating are essential to have under some form of control. Many of the enhancements I have described can be considered as unessential or even frivolous, but I categorize them as convenience. In the balance of this chapter, I will highlight what I consider to be the essentials, and acknowledge those ideas which are clearly related to convenience.

Essentials

The most important essential from my experience is water leak detection. Home automation systems provide a lot of flexibility to protect your home and minimize damages. Water leak sensors are inexpensive and can be placed in multiple locations where leaks are most likely to develop. Connecting the sensors to your home automation hubs can trigger alerts to as many people as you want. The trigger can also close the water main valve immediately to stop water flow. Water leaks are not life threatening, but can cause a disaster which is very costly, time consuming, and inconvenient. A modest investment in leak sensors is a small price to pay compared to hiring mold remediation services.

A second essential is applying home attitude for you and your home's security. Arming your security system protects doors and windows, but the doors do not have to be locked to be protected. With home automation, you can keep your doors locked so no one can walk in when the security is not armed. You could decide to have digital door locks go to the locked position every hour on the hour, at sunset, during lunch hour or dinner, or based on other triggers. I have my hub set to lock the doors at sunset. If you should forget to arm your security system, you at least would want your doors to be locked.

The typical way of arming a security system is to key in a four-digit code and press an Arm button. This works well if you remember to do it. Adding home attitude can help you be sure your system gets armed. You can setup a schedule to have your home automation hub arm your security system every night at a specified time. This may work for some, but the disadvantage of this approach is your evening schedule may have an exception. You may decide to stay up later than normal to read in the family room. The security system gets armed automatically while you are reading. You get up from reading to go to bed, and the motion detector sees you and triggers the burglar alarm. The approach I use and advocate is to have a discrete action, such as pressing the Good Night button on the remote on your night table.

One of my favorite home automation devices is the MiniMote. See figure 12. The MiniMote has four buttons. Button 1 is my Good Morning button and button 2 is my Good Night Button. When I push the Good Night button, the hub turns off the music throughout the home, turns out the lights after a slight delay, and then

Figure 12. MiniMote. By Aeon Labs.

arms the security system. Because the Good Night action list does other things I want to happen before going to sleep, I am very unlikely to forget to push the button. The by-product is an armed security system.

The other essential I recommend is general enhancement of your security system coverage. The security companies do a good job of recommending what protection to put where, but nobody knows your home like you do. Over time, you get to know it even better, and you may give thought to how you can make your security stronger. It might be surveillance cameras in places you did not originally think of, or more motion detectors in places immune from the existing detectors. Security systems have certain limitations to how many security zones you can have. With a strong home attitude, there are no limitations. You can create virtual zones on outdoor sheds, areas of an attic or basement, or any nook or cranny.

In addition to beefing up your detection, you can strengthen the actions which take place when there is a breach. You can make lights blink, send customized messages to friends, family, or property managers. You can use your home audio system to play the loudest most obnoxious music throughout the home, and issue stark warnings to an intruder.

Conveniences

Making things convenient makes them less difficult or accomplished with less effort. Things which are convenient can be useful, not essential, but easy, or even comforting or fun. If a home automation system which handles the essentials can also provide conveniences, the incremental value can be worth the incremental cost. Perhaps not for everyone, but for an increasing number of consumers.

Using home automation to make things convenient is a very personal choice. One person may want a smart flower pot, another person may want the weather forecast broadcast when they push their Good Morning button. In the following paragraphs, I will provide some examples of how a home attitude can provide some conveniences.

Rooms and Lights

The first area of convenience involves rooms and lights. Affordable LED lights are very efficient, but this does not mean you want them on all the time. You can turn them on or off with traditional switches at each lamp or wall switch, but there are many more convenient home attitude methods to use. The simplest is to have a motion sensor in each room. As you enter the room, the lights turn on, and then turn off after 15 minutes or whatever time you specify. Opening a door can have the same effect. If you come into the mudroom from the garage, the mudroom lights come on. After a set number of minutes, the lights would turn off. It could be five minutes for a closet or mudroom, 10 minutes for a powder room, or an hour for other rooms. The times would be based on your experience and preference.

Further convenience can be obtained by using scenes. For example, walking into the family room could turn on the ceiling lights and table lights. Walking into the kitchen could turn on the ceiling lights and set the under-the-counter lights to 50%. When it is time to sit down to dinner, a tap on an iPhone, a wall keypad, or a MiniMote could dim the ceiling lights, set over-the-table lights at 50%, and turn off lights in adjacent rooms. None of these lighting scenes or actions are essentials, they are conveniences. Over time, you can tailor them to your habits.

Perhaps the ultimate lighting convenience at my home occurs at Christmas time. For decades, my wife has meticulously placed holiday wreathes in all the windows of the home which face the street. Each wreath has a small lightbulb. As it gets dark, she goes to each wreath and screws in the lightbulb, and then, at bedtime, she goes back and unscrews the lightbulb to turn it off (easier than unplugging). When the home became a smart home, the process changed. The home automation hub has a schedule for Christmas lights. Each day, between December 10 and January 2, at 15 minutes before sunset, the receptacles where the wreathes are plugged in automatically turn on. At 11:30 PM, all the wreathe lights turn off.

Under the counter, near the kitchen table where we normally have dinner, is a wine cooler. It is a nice thing to have, but the compressor is noisy. It is not noticeable during the day, but after sitting down to a quiet dinner, it can be annoying. The initial solution was to just turn it off during dinner. The problem became remembering to turn it back on. Applying some home attitude solved the problem. When we start the Dinner action group, the various lights are adjusted, the dinner music channel comes on, and the wine cooler turns off. After one hour, it automatically turns back on.

Room Cleaning Robots

An emerging part of home automation includes the use of home robots. Bilal Athar, CEO at Wifigen LLC and a home automation enthusiast, believes robots will be central to having a smart home. "A smart home should also be a clean home.", says Athar. ECOVACS ROBOTICS is a company specializing in research and development, design, manufacture, and sales of robotic home appliances. Their mantra is "Live Smart. Enjoy Life."[88] The company makes robots which add a lot of convenience for cleaning floors and windows. One product, the DEEBOT M81, is designed to clean different kinds of messes in the home. See figure 13 for a picture of the DEEBOT.

Figure 13. DEEBOT by ECOVACS ROBOTICS

The M81 vacuum and mop combo can sweep, vacuum, and mop in one pass. The company says the robot can "give your home a thorough and deep clean." You can choose the cleaning mode to auto mode for general cleaning, edge mode for cleaning specific edges, or spot cleaning when intensive cleaning is required. Everything is controllable from your smartphone. When battery power gets low, the DEEBOT automatically returns to its charging dock. No human intervention is required. You may add some integration with voice assistants like Alexa. A use case could be a child spilled something, and you say "Alexa, ask DEEBOT to clean the kitchen floor".

Music and Announcements

Music is central to the lifestyle of many people. Convenience can play a major role in making the listening easier and more enjoyable. The same is true for home audio announcements. "Whole home" music systems, such as Sonos, can include a speaker in every room. Each speaker is independent, and can play a different music source and have a different volume setting. Some advanced systems can optimize the sound to the acoustical characteristics of the room itself.

My day starts out with a push of button one on the MiniMote on my night table. The audio begins with a good morning announcement of the day, date, time, current weather conditions, and a short-term weather forecast. Then my favorite music plays. The music source is selected at random from a list of a dozen music streaming channels I enjoy most.

Other conveniences include using triggers to announce when something happens. For example, you can have a short audio announcement throughout the home when a door opens, or a thermostat changed a set point, or a leak was detected. The conveniences you can implement are limited only by your imagination.

Summary

The goal for part 1 of *Home Attitude* was to introduce the concepts behind home automation and home attitude. The specific objectives were to explain what home automation is and what benefits it can offer to homeowners. I hope you are beginning to feel a home attitude and are interested in learning more about home automation and pursuing the benefits.

There are three approaches one can take. First is to outline the specific features and functions you would like to have, and hire a systems integrator expert to develop a custom solution for you including everything you want. A second approach is to subscribe to a home automation service from a communications or security services provider. The third approach is do-it-yourself. From here on, I will refer to do-it-yourself as DIY. Each of the three approaches to deploying a home attitude has pros and cons. I will discuss this in more detail in the last chapter. Before moving on to more details about home automation and how to get started, I will discuss the important subject of home automation security and privacy.

CHAPTER 8

Security and Privacy

When your computer, tablet, and smartphone can connect to millions of other computers, hubs, and devices via the Internet, the connectivity facilitates a broad range of useful functions. These include education, e-commerce, communications, collaboration, entertainment, and, of course, home automation. The connectivity with millions of other computers can also make some very bad things possible: loss of your privacy, data, identity, credentials to access your bank account, and electronic health records. Your history of accessing websites and a log of all the actions which have occurred in your smart home can also be exposed.

As described earlier, the benefits of home automation are significant, but the risks of not having a reliable security system can be even more significant. It may not be likely, but it is possible strangers could determine if anyone is at your home, deactivate your security system, open the garage door, back in a truck, fill it with your belongings, close the door, and activate your security system. It would look like nothing ever happened – until you got home.

One way to prevent the home break-in scenario is to not connect your home automation system to the Internet. Criminals would then be limited to the old-fashioned break-in techniques of chopping telephone lines from the security system and lock picking to break in and enter your home. The problem with this solution is you will not be able to use the Internet to connect to your home either. The ability to unlock a door for a maintenance person or change the settings of your thermostats remotely, and other benefits, would not be available. Fortunately, there are techniques available to have it both ways.

The Internet

In order to understand the Internet security issue, it is helpful to understand how the Internet works. All information which travels across the Internet is split into packets. Every email, web page, instant message, tweet, FaceTime video chat, Internet Protocol (IP) telephone call, or remote home automation interaction is broken up into packets which then traverse the Internet. An average packet contains between five and ten thousand zeroes and ones, or bits. The structure of these packets consists of headers and payload. The headers contain information such as the packet's source and destination addresses while the payload contains the actual data being transmitted. The packets move across the Internet by traveling between specialized computers called routers. The packets may not all take the same route from source to destination, but they all end up at the same destination. The routers look at the headers in each packet and determine where it should go next. Typically, a packet may take ten to fifteen hops from one router to another before it gets to its destination. Then, the packets get reassembled into an email, web page, instant message, tweet, FaceTime video chat, telephone call, or home automation request.

Security in all these transactions is critical and needs to be taken seriously because the Internet itself is completely insecure. It is similar to a 1950's party telephone line where multiple parties were actually sharing the same network. Since only one person could make a call at a time, you might pick up a party phone line and find out your neighbor was already using it. Then, if you could overcome the temptation to listen in, you had to wait your turn to use it. The Internet also is a shared network. As our emails, web pages, IP telephone calls, and home automation information travel from origin to destination, a clever snooper could use various tools to inspect or "sniff" the packets to see the contents of the payload. If they are very inventive, they could potentially make changes to the data in the packets. Fortunately, there are many tools and techniques which can be applied to the Internet to make it secure enough for our many uses.

The key to making the Internet more secure is encryption technology, one of the most powerful technologies ever devised. The most basic approach to using encryption is to use ciphers to scramble messages. The concept of encryption has been around for thousands of years. One of the simplest examples of a substitution cipher is the Caesar cipher, which is said to have been used by Julius Caesar to communicate with his army. Caesar used a very basic cipher technique of shifting each letter in a message.[89] The unencrypted message, called plaintext, uses the alphabet. The encrypted version of the message, called ciphertext, is the alphabet shifted 19 characters to the right. An A would be replaced with a T, a B with a U, etc. See figure 14.

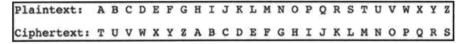

Figure 14. Ciphertext alphabet with a shift of 19

Caesar informed his generals in person of the shift so they could decipher his messages. Even if the enemy intercepted one of Caesar's messages, it would be useless, since only Caesar's generals could read it. An example of a message with a 19-character shift is THE FAULT, DEAR BRUTUS, LIES NOT IN OUR STARS BUT IN OURSELVES. The message would be enciphered as MAX YTNEM, WXTK UKNMNL, EBXL GHM BG HNK LMTKL UNM BG HNKLXEOXL.

Ciphers are used today, although they are exponentially more sophisticated than what the Romans used. Using complex mathematics, the contents of payloads can be scrambled, or encrypted, in such a way only the intended recipient is able to unscramble, or decrypt, the information being sent. Millions of people recognize this technique enables them to put their credit card number into a secure web transaction in a way only the server at the other end is able to read it. People are realizing their credit card number may be safer on the Internet than it is if they provide it to a stranger over a toll-free number for a catalog purchase or to a waiter in a restaurant.

A technology called SSL (Secure Sockets Layer) is a protocol which has been securing web transactions using modern day ciphers for more than 20 years. More recently it has been superseded by the much more secure TLS (Transport Layer Security) protocol though the umbrella term for encrypted communications technology is still referred to as SSL. The concept is to make communications between your web browser and a web server secure so a hacker or snooper somewhere in between cannot read what you send the server or what the server sends to you.

Authentication

There was a cartoon by Peter Steiner in the July 5, 1993, issue of The New Yorker showing a dog at a PC speaking to another dog watching from the floor.

The caption was, "On the Internet nobody knows you're a dog."[90] There are numerous ways to provide authentic identification. I will discuss three: passwords, two-factor authentication, and biometric authentication.

Passwords

Today we use the username and password as the most basic and common form of authentication. We all use them every day, but the problems with them are nontrivial. First is the password sharing problem which enables someone else to be you. If you leave your password on a Post-it® on your PC or under your mouse pad, one of your children or a colleague can become you. They can get into your bank account, buy a book at Amazon, engage in a chat session as you, or reconfigure your smart home to do something different than you want.

Koshevoy Dmitry, author of a website about rules and tips for creating strong and secure passwords, wrote, "The most common password is the word 'password.'"[91] A server connected to the Internet must follow strict security policies including the requirement for strong passwords. Make a New Year's Resolution to get a password manager and clean up and organize all your passwords. You will sleep better and avoid being hacked. There are many different password managers available.

I recommend 1Password, mentioned earlier. It will create, store, remember when needed, and insert hundreds of long, ugly, and impossible to remember passwords. The password to login to my bank account, before I most recently changed it, was

WXXx+mXa>Dq9m@9mnJT%E4ajhm9LuM.

I don't think anyone would be able to guess it. 1Password is a truly great piece of software. It will identify any passwords which are old and those which are duplicates. It is best to have no duplicate passwords. If one of the sites you use gets hacked and the bad guys get your password, you want to be sure it can't be used at any other website. It is easier to have a simple password like abc123 or something equally simple. If you can remember it, it is not secure. Please put getting a password manager high on your list of resolutions.

2FA

Another approach to increasing the security of messages is the use of two-factor authentication (2FA). 2FA requires an additional piece of information in addition to a username and password. The username and password in a single factor authentication is information you know. 2FA supplements what you know with what you have. The things you can have for 2FA include a key fob carried in your pocket, a digital certificate installed on your computer, or an app such as Google Authenticator. After entering a username and password, you are prompted to enter a code from one of the sources listed.

Charles Schwab, a provider of securities, brokerage, banking, money management, and financial advisory services, offers customers a key fob containing a small LCD display. To log in securely to your Schwab account, you enter your username and password (what you know), but then the login page asks for the number displayed on the key fob (what you have). The key fob number changes every 60 seconds. After you enter the number, the Schwab server confirms the number is what it expected based on your identity.

An alternative to a key fob is a smartphone. On my iPhone, I have the Google Authenticator. Authenticator can be used with multiple apps and it generates a random number the apps expect. I use Authenticator with a number of apps including my Gmail account, the Coinbase digital wallet for Bitcoin, Evernote, and my WordPress blog. With two-factor authentication, and a strict password policy, the likelihood of an imposter gaining access to one of your accounts or your smart home is highly unlikely.

Biometric Authentication

A very powerful technique for authentication uses statistical analysis of biological data such as fingerprints or face scans. Examples of biometric authentication using fingerprint recognition are Apple's Touch ID, and similar technology from Google and Samsung. An iPhone user can train the Touch ID software to recognize up to ten fingers. Once trained, the iPhone can recognize you with a simple touch of a finger. Apple explains the security of Touch ID as follows,

Every fingerprint is unique, so it is rare that even a small section of two separate fingerprints are alike enough to register as a match for Touch ID.

The probability of this happening is 1 in 50,000 for one enrolled finger. This is much better than the 1 in 10,000 odds of guessing a typical 4-digit passcode. Although some passcodes, like "1234", may be more easily guessed, there is no such thing as an easily guessable fingerprint pattern. Instead, the 1 in 50,000 probability means it requires trying up to 50,000 different fingerprints until potentially finding a random match. But Touch ID only allows five unsuccessful fingerprint match attempts before you must enter your passcode, and you can't proceed until providing it.[92]

Touch ID doesn't store any images of your fingerprint. It stores only a mathematical representation of your fingerprint. It isn't possible for someone to reverse engineer your actual fingerprint image from this mathematical representation. The chip in your device also includes an advanced security architecture called the Secure Enclave which was developed to protect passcode and fingerprint data. Fingerprint data is encrypted and protected with a key available only to the Secure Enclave. Fingerprint data is used only by the Secure Enclave to verify that your fingerprint matches the enrolled fingerprint data. The Secure Enclave is walled off from the rest of the chip and the rest of iOS. Therefore, iOS and other apps never access your fingerprint data, it's never stored on Apple servers, and it's never backed up to iCloud or anywhere else. Only Touch ID uses it, and it can't be used to match against other fingerprint databases.[93]

Apple has made the Touch ID application programming interface (API) available to developers. The result is many apps, including one from my bank, accept my fingerprint as a secure method of authentication. The iPhone, iPad, and some models of the MacBook support Touch ID. If your phone is protected, only you can access your smart home app. Android phones use various similar implementations of fingerprint scanner technology.

An emerging new method of authentication is 3-D scanning of our faces. The tremendous power of the smartphone will make this biometric technique possible and the capability will evolve as we have seen with audio and video. In late August 2017, Brian Chen wrote in *The New York Times*, "As soon as you pick up your gadget, it will see you and know you are the owner and unlock the screen."[94]

On September 12, Apple announced Face ID. The company said, "With a simple glance, Face ID securely unlocks your iPhone X."[95] With Face ID, you can make purchases from Apple and make payments with Apple Pay. Apps which support Touch ID will automatically support Face ID, and many more will surely follow. The technology behind Face ID is quite impressive. The camera captures and analyzes more than 30,000 invisible dots projected onto your face plus an infrared image of your face. The data is compared to the data from when you enrolled yourself. Face ID works in the dark and can adapt to shaving a beard, and wearing a hat, scarf, or sunglasses. While the odds of someone stealing your iPhone and having a fingerprint just like yours is 1 in 50,000, with Face ID, the odds of the thief having a face the same as yours is 1 in a million.

Strong passwords, 2FA, and biometrics add greatly to strengthening security. However, in addition to the concern about accurate authentication, there are other cybersecurity issues. These include breaches and distributed denial of service attacks.

Breaches

If you choose the subscription model for home automation, the question is whether the security or telecommunications company you select has a secure infrastructure. It is a shared responsibility. You need to select a strong password and change it at least twice per year. The other half of the responsibility is for your home automation provider to protect against breaches of their servers.

Breaches of servers connected to the Internet have emerged as a significant global security concern. Jake Kouns, co-founder and President of the Open Security Foundation which oversees the operations of the Open Source Vulnerability Database, tracks data breaches in his DatalossDB.org blog. He reported 2015 had an all-time high 3,930 breaches of servers which exposed more than 736 million records. Forty-one percent of the servers and sixty-five percent of the records were in the U.S.[96] Kouns said, "Email addresses, passwords and usernames were exposed in 38% of reported incidents. Passwords were the most sought data to steal." Kouns said, "This is especially troubling since a high percentage of users pick a single password and use it on all their accounts both personal and work related." Weak passwords represent a security risk, but there are numerous tools available to create and manage strong passwords.

Clearly, a strong password and 2FA are essential to prevent a brute force attack against an online account. However, even with 2FA it is possible to breach a systems' security if SSL is improperly configured by your provider, or if the provider does not keep software up to date. The now infamous 2017 data breach at Equifax, exposing the data of 143 million people, makes this clear. The company failed to apply a publically known security patch to its server software which could have prevented the breach.[97]

Hackers understand encryption technology. There have been a number of high-profile security breaches caused by improperly configured SSL based web servers. For example, a known vulnerability in the older SSL protocol allows hackers to break the encryption which otherwise would protect sensitive data. For this reason, it is imperative for organizations to be vigilant with best practices, such as enforcing the newer TLS protocol and disabling the legacy protocols altogether. Security best practices also include regularly applying security updates to prevent susceptibility to emerging threats when they are discovered.

A study of 2015 breaches was done by the Online Trust Alliance, a Bellevue, Washington charitable organization with the mission to enhance online trust and promote innovation and the vitality of the Internet. Its results were released in January 2016. It reported 91% of the data breaches occurring from January to August of 2015 could have been easily prevented using simple and well-established security practices.[98] The possibility of breaches should not be ignored by your provider. They are manageable if good security practices are followed.

Distributed Denial of Service Attacks

One of the most feared cybersecurity threats is from a Distributed Denial of Service Attack (DDoS). DDoS attacks are typically associated with a Trojan horse virus. The term Trojan horse comes from a Greek tale about the Trojan War.[99] The Greeks built a giant wooden horse and presented it to the Trojans, making it look like a gift. After the Trojans dragged the horse inside their city walls, Greek soldiers exited the horse's hollow belly and opened the city gates. Greek soldiers were then able to swarm in and capture Troy. The hackers' Trojan horse is made from software, not wood. The software is a virus, meaning it can have a detrimental effect on a computer such as erasing data or performing inappropriate actions.

DDoS is a coordinated attack in which many compromised systems, typically infected with a Trojan horse virus, are remotely commanded to repeatedly make requests to a single system, thus overwhelming the system's resources and causing a denial of service to legitimate users. If a successful DDoS attack is perpetrated on your home automation provider, you would not be able to access your home automation system.

Fortunately, it is possible for a provider to plan ahead and thwart DDoS attacks. There are two basic approaches to protecting against a DDoS attack. One method is to add significant capacity. If the capacity is present, the attack has no consequence. All of the login attempts are handled normally, and only the legitimate ones are allowed access to the home automation server.

The second defense to a DDoS attack is to prevent illegitimate requests from infected systems getting into the home automation server. This can be done with specialized hardware devices which are connected to the Internet and to the home automation server. The hardware device acts as a first line of defense. It inspects the contents of the incoming message and determines if it is legitimate based on where it came from and how the request is structured. Numerous vendors, including Akamai, Arbor Networks, Barracuda, Cisco, Radware, and Verizon have sophisticated solutions specifically designed to protect against DDoS attacks. The Arizona Democratic Party in March 2000 experienced two types of cybersecurity attacks, brute force password guessing attempts and DDoS. Both attacks were thwarted successfully because the proper protections were in place.

The Tunnel

If you choose the DIY route, security should play a key role. Indigo and other home automation systems offer various security options. The absolute minimum you should have is a strong password to your hub. For my system, I chose to use a virtual private network (VPN), often called a tunnel. A VPN tunnel uses the Internet, but it creates a virtual tunnel for communications between the computer running your hub and your smartphone. See figure 15 for a pictorial diagram.

All exchanges of data between the hub and the smartphone are encrypted. When I connect remotely to the Mac, which is running Indigo, I need the login and password for the Mac, but I also need a Shared Secret Key to establish the secure tunnel. The line below shows my secret key, before I last changed it. The key should be something no human could guess or remember.

j$tTFPkn(bN=VBTvNY@Kx8tBuamhE>]q8WXT2bJT3Mcov4vpxZ)
Bo7uMB9$g4Mre

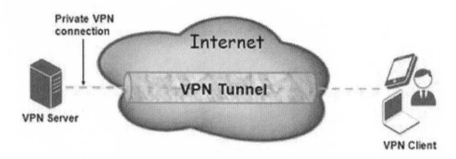

Figure 15. VPN Tunnel from D-Link Business Blog

The steps to connect are simple. Using an iPhone, I open Settings and select VPN. I then click Connect. My user id, password, and shared secret are all stored in the settings on the iPhone. If someone steals my iPhone, he or she will need my fingerprint. Once I click Connect, and the iPhone shows "Connected", I am able to use the tunnel which was created. The tunnel, in effect, makes my iPhone a device on the local area network (LAN) in my house. In other words, the secure tunnel over the Internet enables me to do anything with the iPhone I could do if I was physically in the house and connected to Wi-Fi. My hub settings ensure only local connections are accepted. There is no way to get to my hub except by a local connection. The VPN tunnel makes me "local" even though I may actually be on a different continent.

VPN software is available for Mac, Windows, and Linux platforms. You can also use a VPN service. The VPN service provider gives you access to them via login and password, and then uses their server to connect to yours.

Privacy

Nashville, Tennessee based consulting company Clearwater Compliance says stolen credit and debit card numbers sell for as little as $1 or less. However, stolen personally identifiable healthcare information is anonymously bought and sold for as much as $60.[100] Fortunately, we have not heard much about theft of home automation data. Yet.

Suppose you have a weak password for the hub you have at home or the service to which you subscribe is not well protected. What could be stolen could be quite valuable to thieves and burglars. The most egregious case would be a thief gaining access to your hub, disabling the security system and video cameras, walking into your house, stealing everything of value, walking out, and resetting the security system to Armed Away. Such an event represents a breach of security and an invasion of privacy.

More subtle infringements are possible. Most hubs have a log. Every time a device status changes, a scheduled event occurs, a trigger takes place, a variable value changes, or any action is executed, an entry is made in the log showing the data and time. With access to the log, an intruder has a treasure trove of data about what goes on in your smart home.

A patient thief may be interested in a more significant booty. By studying the log over a period of time, the would-be thief might discover a homeowner is consistently away from home with the Arm Away status set the first weekend of each month. For this more nefarious operation, the thief could disarm security, open the garage door at sunset, back in a truck, and take all the time needed to fill it.

Some smart homeowners use a home automation plugin to control the use of their Sonos wireless home sound system. The log would show every selection of music. The homeowner may be very private and perceived as an exclusive listener of fine classical music. The intruder learned from the log the homeowner is actually an avid fan of risqué rap music. The homeowner might receive an anonymous email showing a list of all the songs selected, how often, and at what times. The email might also have instructions to scan the QR code which follows and send $5,000 in bitcoin ransom to the perpetrator. The email might specify failure to do so within 24 hours will result in the music log being sent to the homeowner's boss and colleagues. Listening to whatever music the homeowner chooses is certainly his or her right, but the revelation would be a major invasion of privacy.

Voice

When you ask Alexa or other voice assistant a question or give her a command, she not only understands it, but saves it. Open the Alexa app on your smartphone and you will see everything you have ever said to her on display. You can delete your questions and commands one at a time in the app or go online to amazon. com and delete them all. Nevertheless, capturing the things you say provides data similar to what your browser does with its history. The more data Amazon, Apple, Google, Microsoft, and Samsung have, the better they know you and your needs and interests. It comes down to trust. Do you trust them? If not, put masking tape over the video camera on your desktop or laptop, and use "Incognito Mode" on your browser.

Many have worried about the government watching us. Now, the technology allows us to watch the government. The tech companies know we are watching them. They know we expect all of our data to be encrypted. If any one of them drops the ball on security or improperly violates our privacy, they know we can switch to a competitor in a few mouse clicks or taps.

Conclusion

Strong security is a critical factor for a successful and trusted home automation system. Concerns about Internet security should not deter you from accessing your smart home remotely. If you have chosen to use a subscription service from a security or telecommunications company, ask them what security procedures they offer and if they have protected themselves from breaches and DDoS attacks. If you select an integrator to build a system for you in your home, ask them to include the strongest security options available. At a minimum, all three approaches to implementing a home automation system must include use of strong passwords, and ideally require two-factor authentication or a secure VPN service. Strong security is a key part of a healthy home attitude and essential to protect your privacy.

If you are a DIYer, you can enjoy adding and changing devices, schedules, action groups, and scenes, and making them all work together to make your home a smart home. If you have taken the subscription or integrator route to home automation, with proper security in place, you can also relax and enjoy the comforts and safety of your smart home.

Part 2
Home Automation
How It Works and
How To Get It

Although there are many approaches, options, and features available when creating a smart home, there are some basic components which comprise home automation. They include devices, hubs, schedules, action groups, plugins, triggers, scenes, and user interfaces. These have all been discussed at a high level in part 1. Part 2 has some overlap with part 1, but will go into more detail. The focus of part 2 is DIY, but if you choose to go the route of hiring a professional integrator or getting subscription service, you may want to learn more details to help you refine your requirements. You may also just be curious about how the components work.

In the following chapters of part 2, I will describe the basics of home automation components, some examples of what is available, and more details about how the components work. It is not possible to cover all the home automation products and servicees available, and many more are coming to market, but part 2 should provide plenty of home attitude ideas on how you can make your home smart.

CHAPTER 9
Hubs and Networks

The central part of a smart home is the home automation hub, which I will refer to simply as a hub. The hub is analogous to other kinds of hubs: the hub of a wheel which connects the spokes, a geographic regional hub which facilitates economic activity, or a transportation hub which expedites national and international transportation. Every smart home has at least one hub which serves as the commander in chief. The hub coordinates commands to and from the devices. A hub, sometimes called a controller or gateway, can be a standalone device or a desktop, laptop, smartphone, tablet, or smart watch. A hub can be as simple as a single purpose hub to control the light on the front porch, or it can be a very sophisticated hub which controls everything in the smart home including hundreds of devices.

The hub uses specialized home automation software to tell devices what to do and when to do it. To make this possible, hubs have special capabilities including schedules, action groups, triggers, and scenes. The nomenclature may differ among the various vendors, but the capabilities are basically the same. Hubs have a user interface which allows smart home owners to use these capabilities. The interface can be a smartphone app, a smart watch, a web page, or your voice.

I will describe two hub alternatives out of the many choices available. First is the Samsung SmartThings Hub, available for $99. It requires an iPhone or Android phone for the user interface. The SmartThings Hub is easy to set up and easy to use, although somewhat limited in its capabilities and compatibility with other devices. The second hub I will describe is the Indigo Smart Home Software platform. Indigo has a very broad set of features and capabilities. It offers the do-it-yourself hobbyist a virtually unlimited home automation solution. Indigo costs $250 and requires an Apple Mac computer.

SmartThings

The SmartThings Hub is made by an independent subsidiary of Samsung Electronics based in Mountain View, California. The hub is white, measures 4.2" x 4.9" x 1.3", and weighs 7.68 ounces.[101] Inside is a computer chip which runs the home automation software and interacts with the SmartThings app on your smartphone.

The hub connects to your home's local area network (LAN) router via an Ethernet cable. The Hub works with devices which support the Zigbee, Z-Wave, or Bluetooth wireless protocols and also supports devices which are connected to your LAN via Wi-Fi. Supported devices include lights, switches, outlets, sensors, cameras, doorbells, door locks, thermostats, and speakers. The hub contains replaceable batteries which allow it to continue operating in the event of a power outage.

As mentioned earlier, the SmartThings hub allows you to monitor and control the connected devices using an iPhone or Android app, or using the Amazon Echo or Google Home voice assistants. The SmartThings Hub can send you alerts from connected devices if there is unexpected activity in your home reported by a motion detector or the opening of a door or window. The hub can turn lights on or off when doors are opened or a person enters a room. Using SmartThings Routines, you can set up actions such as Good Morning, Goodbye, or Good Night. Routines can include as many actions as you want. The hub also supports schedules. You can have a routine take place every day at a certain time or at sunset or sunrise. You can also specify a routine to take place only on certain days of the week. Devices connected to the hub can trigger an event. For example, if a door opens or motion is detected in the kitchen, lights can turn on or you can receive a notice via email.

All things considered, the Samsung SmartThings hub is a good choice for anyone wanting to make his or her home smart with a minimum of effort. No technical knowledge is required. The tradeoff is simplicity versus full-scale advanced home automation. The limitations are twofold. First is the devices. As of August 2017, the smartthings.com website shows 142 supported products. For most people, the categories and choices are adequate, but if there is a special device, such as a propane tank sensor, a more advanced system is required. There are hundreds of devices which are not supported. The second limitation is in the level of granularity in the actions and schedules. For example, if you want a porch light to turn on 20 minutes after sunset instead of at sunset, a more granular capability is needed. SmartThings support is available in the traditional way by phone, email, and live chat.

Indigo

The Indigo Smart Home Software platform is a product of Dallas, Texas based Perceptive Automation, LLC.[102] Unlike SmartThings, Indigo does not include any hardware, it is 100% software. The Indigo software runs on an Apple Mac computer. It does not have to be a dedicated Mac. I wrote this book with Microsoft Word running on an Apple iMac. The iMac also runs Indigo. Some Indigo users choose to use a dedicated Mac where nothing is running except Indigo. The Mac Mini is a popular choice for the dedicated option because of its attractive price, especially if you buy a used one. To get started, you download and install Indigo on the Mac, just like you would for any other Mac app.

The Indigo software is very easy to use. It takes advantage of all the elegant user interface design elements for which Apple is well known. Indigo can be used for simple home automation tasks, but also has highly advanced capabilities for those who need it. Indigo supports hundreds of different smart home devices, and if there is a device you want to use which is not supported, there is likely a 3rd party plugin which will provide the support.

Indigo has many advanced features. For example, you can easily design graphical control pages for your smartphone which are specific to your home. Comprehensive scheduling and triggering features support very complex logic which can enable you to go far beyond the basics and create a truly intelligent smart home automation system. Indigo includes an intuitive user interface which allows you to take advantage of advanced features without having to know exactly they work. If you have technical skills and some very advanced ideas, you can use a form of computer programming, called scripting, to add advanced functions. With Indigo, the possibilities are unlimited.

Excellent support for Indigo is provided through a How-To Wiki, User Forums, and an extensive library of useful tools and files. The developers of Indigo are actively engaged in helping users, but the users themselves are at least as valuable in resolving problems. They walk in your shoes. The user forum for those who are new to Indigo or home automation has more than 2,000 posts which have been shared.

Indigo's biggest strength is the extensive community of users who not only help each other out on the user forum, but have also created some very high-quality plugins which extend Indigo's usefulness by supporting hundreds of additional devices and protocols.

One advanced feature which Indigo offers in a unique way is the use of variables. For example, you can create a variable named home_status. The variable may be equal to "home" or "away", and actions can take place or not based on this value. If the value is "home", the solar shades will go up in the morning. If the variable equals "away", the shades will not go up.

Start Simple, Upgrade Next

If you don't have a Mac and are not ready to take the plunge, you can start simple and upgrade later. For example, you can start with a SmartThings Hub and add the lights, switches, thermostats, door locks, and other devices you want to get started. If you later decide to get a Mac and install Indigo, it would be very easy to migrate all your devices from the SmartThings Hub to the Indigo software platform.

Networking

Hubs communicate with devices via networks. A simple example of a network is a cordless phone. Although in decline, there are millions of them. A cordless phone has a base station which is connected to your landline and telephone service provider. When not in the base station, a portable handset communicates with the base station via a wireless network. Communications between the base and the phone takes place using a network protocol, often called a standard. There have been more than a few different standards for cordless phones over the years. Panasonic, the market share leader, currently uses Digital Enhanced Cordless Telecommunications (DECT).

Similarly, home automation devices communicate with a hub using a network. See figure 16 for a pictorial sketch of a hub with networks and devices. Applying a home attitude would be much simpler if there was one network protocol used by all devices and hubs. I have heard many people say we need a standard. There is no lack of standards. Unfortunately, there are many of them, and they are not compatible.

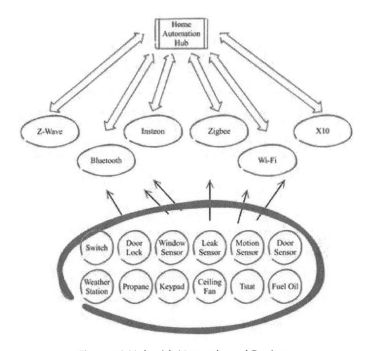

Figure 16. Hub with Networks and Devices

The Internet

Contrast the world of home automation with the Internet. The Internet provides a ubiquitous communications network. TCP/IP (the transmission control and Internet protocols) enabled global networks to become interoperable. I discussed Internet standards and their significance in more detail in *Net Attitude: What it is, How to Get it, and Why it is More Important Than Ever*.[103] It is not necessary to understand the Internet standards as part of your home attitude. The important thing to know is the Internet works exactly the same way everywhere in the world.

The lack of a single network standard is the biggest challenge in home automation. The most common home automation network standards include Bluetooth,

Insteon, Wi-Fi, X10, Z-Wave, and Zigbee. Compatibility among devices and hubs is a mixed bag. If your hub supports only Z-Wave, and your door lock uses Zigbee, your hub will not be able to lock and unlock your door. The Samsung SmartThings Hub supports Bluetooth, Wi-Fi, Zigbee, and Z-Wave protocols. It does not support Smarthome's Insteon protocol. Smarthome's hubs do not support Zigbee or Z-wave. Indigo supports Z-Wave, Insteon, and X10 network protocols. These protocols are built into the Indigo software platform. Some of the network protocols require their own hub, but Indigo integrates them. You can mix and match any combination of devices with those protocols. With more than 200 software plugins which extend the Indigo software platform, there is no limit to what can be supported.

Each of the network protocols has pros and cons. Some have longer wireless ranges. Others are said to be more reliable. The biggest variable is often the number of walls and obstructions the wireless signals have to travel. The best strategy is to read about the experience of others on the user forums and wikis. My devices use Bluetooth, Wi-Fi, or Z-Wave. Indigo makes them work together nicely. I have not experienced any problems with interference or range.

Summary

The hub and one or more network standards provide the backbone of your smart home. The two hubs described show the range of capability available. There are many other choices you can consider. Make a list of the devices you want to have and make sure whatever hub you select can support all of them. Now we will take closer look at some of the devices you can choose from.

CHAPTER 10
Devices and Plugins

A device is a thing, usually something mechanical or electronic. You could say a hammer is a device, or a mouse trap. However, in the world of home automation, devices are smart devices. A smart device is an electronic device which is generally connected to or communicates with a hub and other smart devices. We can interact with smart devices, or in some cases the devices can act autonomously, make decisions, and take action without us making requests. Over time, smart devices will continue to get smarter and develop very powerful artificial intelligence.

Devices form the basic building blocks for a home automation system. Smart devices enable you to have a smart home. Smart home devices have a computer chip in them which makes them smart and enables them to communicate, either via a wire or wirelessly. The devices can communicate with other devices or with a home automation hub, which acts like a central point of contact for all the smart devices.

Devices come in many sizes and shapes, and can perform many different functions. A simple example is a smart lightbulb. The bulb has a chip in it. When it receives a command from a device or hub, it will turn itself on. A different command causes it to turn off. As mentioned earlier, some more sophisticated lightbulbs can also respond to commands to dim or brighten. Some even smarter bulbs can change the color of the light they emit.

Chips

Smart devices gain their smarts from a computer chip built in to them. The first chip was invented in 1958 by Jack St. Clair Kilby, an American electrical engineer who worked at Texas Instruments (TI).[104] Although the TI chip was not a computer chip, it contained integrated circuits, and it revolutionized the electronics industry. The first chip with a computer, called a microprocessor, built into it was the Intel 4004. It was invented in 1971 by Intel engineers Ted Hoff, Federico Faggin, and Stan Mazor. The Intel 4004 was smaller than a thumbnail, contained 2,300 transistors, and could execute 60,000 operations per second.[105] By comparison to chips available today, the 4004 was not very powerful. Apple's A11 Bionic chip, introduced for the iPhone 8 and X in September 2017, contains 4.3 billion transistors.[106] Many of today's chips can perform hundreds of billions of operations per second. It has six cores each of which can perform specific types of processing, and can perform 600 billion operations per second.

Chips embedded in various devices give the devices amazing power and capability. The Apple Watch has GPS built in. Tiny communications chips enable door locks, thermostats, and light switches to communicate with hubs. Many devices can encrypt and decrypt the data they send and receive. Most new printers offer built-in Wi-Fi chips. Microprocessor chips, such as the Apple A11, make handheld smartphones vastly more powerful than most desktop computers in use.

Device Types

Devices for lighting include switches, keypads, dimmers, bulbs, and receptacles. Entertainment devices include audio speakers, flat panel displays and TVs, remote controls, and devices for streaming of audio and video content. Sensor devices can detect sunlight, ultraviolet temperature, humidity, water leaks, motion, and vibration. Devices with motors in them include ceiling fans and solar shades. Smart thermostats can respond to commands to change heat and cooling set points, HVAC mode, and fan mode. Specialized devices can open or close a water main valve, start an irrigation system, or communicate with and control other devices. See table 10 for a partial list of device and sensor types.

LED Strip

One new and highly innovative device worthy of special note is the LED Strip from Aeotec.[107] It can change how you think about lighting. Like all Z-wave connected lights, the LED Strip can turn on when you enter a room or turn off when you're not home, but it can do much more. The LED Strip is 16 feet long, slightly

Table 10. Device and Sensor Types

Device Types	Sensor Types
Audio/Video Streamers	Door
Ceiling Fans	Humidity
Dimmers	Luminance
Door Locks	Motion
Energy Monitors	Oil Level
Garage Door Controllers	Pressure
Hubs	Propane Level
Irrigation Controllers	Rain
Keypads	Snow
Lights	Tampering
Receptacles	Temperature
Remotes	Ultraviolet
Sensors	Vibration
Switches	Water Leak
Thermostats	Water Level
TVs	Weight
Weather Stations	Window

less than a half-inch wide, and an eighth-inch thick. In approximately every eight inches of the strip, there are six semiconductor chips which emit cool white light, six which emit warm light, and six which can emit any of 16 million colors. The back of the strip has an adhesive stick-on surface, so the strip can be placed anywhere, including outdoors.

The possibilities are endless. Lighting can now be decorative. You can light your rooms, halls, and stairs with perfect colors. You can make the light bright or soft, or you can make it change based on conditions. For example, using IFTTT described in chapter 2, you can have your kitchen lights flash green when the Domino's pizza delivery car is about to pull up in your driveway.[108] You could have the room blink blue when an email is received, or have the lighting flash red if there is a water leak or other problem. When having a party, the lighting could be set to alternate to a mix of colors. If not already, there probably will be a plugin to synchronize the lighting changes with the music playing on your home audio system.

Availability

A smart home can have one device, such as a digital door lock which locks or unlocks from your smartphone, or it can have hundreds of devices all coordinated from a hub. AliExpress, Amazon, and eBay list tens of thousands of home automation devices. You can buy a Z-Wave Water Leak Detector at AliExpress for $25 with free shipping, or a HomeSeer Home Troller S6 PRO Home Automation Controller for $1,199 at Amazon. You can set up many devices yourself, but for devices which require 110-volt wiring, it is best to hire a licensed electrician. Smart door locks can be tricky to install and align properly. Experienced carpenters know how to it. Smart thermostats can also be challenging to install. I have learned this the hard way. An experienced HVAC technician knows how to swap out the old thermostat and connect the new one without a problem.

Plugins

Computer software applications are each designed to perform certain tasks. Browser programs like Chrome, Firefox, or Safari are designed to browse the World Wide Web. Microsoft Word is designed to create and edit text documents. HomeSeer and Indigo are designed to perform home automation tasks. In many cases, the user of a software app may want the app to do something it was not designed to do. Many software apps are designed to allow their functionality to be extended with a plugin which adds the missing function the user wants.

A plugin, sometimes called an add-in, add-on, or extension, is a software component which adds a specific feature to an existing computer program. In the case of Chrome, an extension I use is the 1Password Password Manager. As the name implies, the extension remembers all my passwords so I do not have to. It inserts them automatically when I login to a website. In Microsoft Word, I use a plugin from Clarivate Analytics called EndNote. The plugin allows me to easily insert footnotes and bibliographic reference information for the books I write.

In the case of home automation software, the most useful kind of plugin is one which adds support for a device which is not normally supported by the software. Many plugins are created by users who need a capability and have no other way to get it. The community of users typically share and support what they create. For example, a home automation enthusiast in the Chicago area who goes by DaveL17 in the Indigo community forum, wrote a plugin which enables Indigo to retrieve weather information from Weather Underground and report weather conditions from the website or local user supported weather stations.

Nick Lagaros, a financial industry technologist and home automation enthusiast has written plugins to allow Honeywell Wi-Fi thermostats and Sonos home audio systems to work with Indigo. I use a plugin which enables me to automate an Epson home theatre projector. Another one enables me to connect my home security system to Indigo and treat each door, window, smoke alarm, etc. as a home automation device. Indigo has been extended with more than 200 plugins.

In my opinion, plugins are an essential part of home automation, especially for DIYers. This is why I have had so much to say about Indigo. It is an outstanding model of how to enable the creativity of technically oriented hobbyists and developers.

Home automation is becoming mainstream. There will be a flood of new smart home devices in the market. Some of them will be embraced immediately and supported by hub providers, but some will not. If there is one you really need or want, you are out of luck, unless a fellow DIYer or the device manufacturer develops a software plugin which enables it to work with your hub.

Maintenance

One final note about devices. They need maintenance, just like non-smart devices. One should not assume, once you have everything setup and working, your smart home will work error free forever. Smart door locks and leak sensors have batteries. Batteries run down. Although it is possible to find a plugin or write a script to automatically check battery levels and alert you when they are low, it is a good idea to have a checklist, like Spring cleanup, to check battery levels. Once you have enough experience to know the life of the batteries, you can establish a schedule for replacement. This is especially important for leak sensors.

Summary

There are thousands of home automation devices to choose from. Although I have provided a number of examples, it is not possible to cover them all. Before you get started, visit the AliExpress, Amazon, eBay, and Smarthome websites and search for "home automation devices". This will enable you to get an idea of what is available, how much they cost, and what hubs they work with. You can also visit the Indigo and Samsung wikis and follow the discussions of users to see what they like and what they may be having trouble with. Next I will discuss what devices can do when they part of actions and scenes.

CHAPTER 11
Actions and Scenes

Actions

The key to getting the most from your smart home is to have it do things which enhance your safety, security, enjoyment, and convenience. The smart home does things when the hub executes an action. Some hubs call doing something a routine. There may be other terms used, but they all mean the same thing. Actions can be to turn a device on or off, change the heat set point of a thermostat, play a music source, change the music volume, or announce the time of day or weather forecast through an audio device (speaker).

An action can be to set the value of a variable. For example, the hub might contain a variable called security_status. The value could be set to "armed" or "disarmed". Another variable, mentioned earlier, might be home_status with values of "home" or "away". If you are away and the security system is armed, an action could put your lighting in a random mode so lights go on and off at random during certain hours to make it look as though you are not away. Another action could be a notification, which can be done by text, email, or a voice announcement. The message could be about a water leak detected in the basement or reporting the weekly backup generator test failed to run and attention is required.

As implied earlier, variables can be used to condition certain other actions or schedules. I have variables which are used to set the smart house thermostats in the morning and evening. The value of the variables changes based on the season. For example, the sunroom and master bedroom thermostats are set based on the value of vaiables:

tstat_sunroom_summer_morning_cool = 72 and
tstat_mbr_winter_night_heat = 65.

An action group is a group of actions which could be executed together. For example, the hub might contain a Good Night action group which includes actions for turning off smart home lights, lowering the volume of music for ten minutes and then turning it off, setting thermostats to a night setting, and locking the doors.

There is no limit to what you can do with a group of actions. For a complex set of actions, I like to use a script. What follows is an example of a script which is executed when I push button 1 on a MiniMote remote control for the Good Morning actions to begin. The script was written using the Python scripting language, but I will show the list of actions in plain language to make it clearer.

- ✓ Start the Good Morning actions.
- ✓ Get the season, date, and time from the hub.
- ✓ Lower the motorized solar shade.
- ✓ Wait 15 seconds and then turn on the night table lamp.
- ✓ Wait 10 minutes and then turn on the ceiling fan light.
- ✓ Wait 15 minutes and then set sunroom thermostat based on the morning temperature variable for the current season.
- ✓ Wait 15 minutes and then unlock all doors.
- ✓ Prepare the good morning announcement.
- ✓ Format the date like Saturday, August 19.
- ✓ Get the temperature and humidity from sensor outside the garage.
- ✓ Round off the temperature and humidity values.
- ✓ Get the latest forecast from Weather Underground plugin.
- ✓ Convert WU forecast: remove F from the temperatures and convert wind direction, like ESE to east south east.
- ✓ Concatenate components of good morning announcement
- ✓ Use Amazon Web Services Polly artificial intelligence to create an announcement using Salli's voice. "Good morning. Today is Saturday, August 19. The time is seven thirty. The temperature outside the garage is seventy degrees and the humidity is 63 percent. Today's forecast is partly cloudy with isolated thunderstorms developing this afternoon. High near 80. Winds west northwest at 5 to 10 miles per hour. The chance of rain is thirty percent. Have a nice day at the Lake House."
- ✓ Use the Sonos plugin to play the announcement from the master bedroom speaker.

✓ Make a music selection at random from a list of favorite stations, playlists, and albums. If it is Sunday, select at random from list of Baroque music sources.

✓ Play the selected music on Sonos speakers throughout the house. Set volume levels based on preference for each room.

It is best to start simple, and add things to your action groups or script as you gain confidence things are working the way you want. DIYers can experiment with new capabilities, and then modify or scrap them as desired. One variation you could make to the Good Morning sequence is to incorporate the Aeotec LED Strip. When the lights come on in the morning, you could set the room color based on the weather forecast: blue for clear skies, grey for cloudy, and yellow for rain or snow.

Action Types

Most actions are related to devices. For lighting, they would include turning a light on or off, dimming or brightening it, toggling lights between on and off, or in the case of color lights such as the LED Strip, setting the color. For door locks, actions can lock them or unlock them. For an irrigation system, actions include running, pausing, or resuming a schedule, activating zones in sequence, or turning on a specific zone. For thermostats, there can be more than a dozen actions including setting heat or cooling set points, setting the system mode and the fan mode, and retrieving the temperature or humidity. For a ceiling fan, the actions would include setting the fan speed to high, medium, or low, increasing or decreasing the fan speed, and turning the fan on or off.

Actions can manipulate variables. For example, a variable can be set to a specific value such as making summer_sunroom_morning equal to 70 as in the Good Morning script. A variable such as shed_door_open could be set to true or false. A variable can also be incremented or decremented by 1. You might have an action called Sunroom Cold. When executed, the action would add 1 to the thermostat heat set point.

A notification action can cause the hub to send an email. These actions provide a way to confirm a particular action had occurred or a certain state achieved. For example, you could have the hub send you an email if a certain temperature had been reached inside, or outside if you have a sensor. Email notifications for alerts are particularly useful when you are not home. You can use various web services to convert an email to a text message.

Most hubs can execute the basic actions of turning things on or off, locking or unlocking a door, or adjusting a thermostat setting. Indigo has approximately 100 types of actions you can select from. In addition, plugins have actions. If you have a home attitude, the possibilities for actions are limitless.

Scenes

If you use Indigo and have a lot of Smarthome's Insteon switches and receptacles, you can simplify actions by using scenes. Scenes are designed to make the automation of the smart home more efficient. For example, if you want to automate lighting in the kitchen at dinner time, you could create an action group containing a separate action to turn on or brighten each light. The kitchen might have counter lights, ceiling lights, kitchen table lights, etc. Each action in the action group would turn one of the lights on or off. An alternative method is to create a scene in Indigo which includes all the kitchen lights, each with its own brightness level. With a scene in place, automating the lights is a simple matter of one action which turns the entire scene on or off.

Summary

Actions and scenes are what makes things happen in your smart home. They can be simple or complex. The best advice is to start simple and add as you get more experience. Next I will discuss how schedules and triggers can make things happen.

CHAPTER 12

Schedules and Triggers

Actions and scenes can do many things, but only if they are executed. This chapter is about making things happen, making your actions and scenes come to life. One simple way to make them execute is by pushing a button on a keypad or remote. Another way is to use your home automation app and tap an icon like Good Night. Increasingly, we will just speak to one of our voice assistants such as Amazon's Alexa, Apple's Siri, Google Home, Microsoft's Cortana, or Samsung's Bixby. "Alexa, turn on the kitchen lights." or "Bixby, good night." The most powerful way to cause an action or scene to execute is to not use any of the aforementioned methods. The power of home automation, and a home attitude, is automation, things happening on their own. The two methods of automation to follow are schedules and triggers.

Schedules

Schedules provide a way to make things happen on a regular basis. For example, a hub may have a simple schedule named Porch Light which turns the porch light on every day at 7 PM and then turns it off at midnight, but schedules can be much more powerful and flexible. The time can be X minutes before or after sunrise or sunset. You may want to use a schedule to make sure you don't miss a recycling pickup which occurs every other Wednesday. You can set up a schedule to execute at 7:00 AM on the first and third Wednesday starting on June

1 and ending on August 31. You could condition the schedule to only happen if the Wednesday is not a holiday. The actions to happen when scheduled would include turning the front lamp post on and making an announcement on your home audio system to remind you to take out the recycling. See figure 17 for a pictorial showing how the schedule could be created in Indigo.

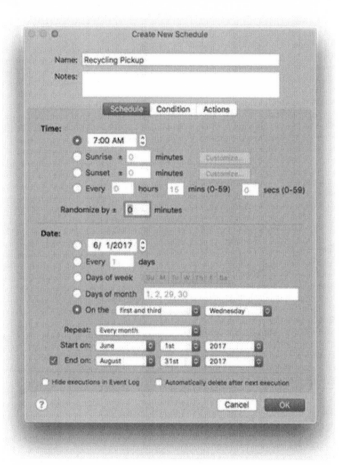

Figure 17. Recycling Pickup Schedule

Times for turning home lights on and off can be plus or minus a random number of minutes within a certain range so the smart home can appear to be occupied even though there is nobody home. A schedule can be executed automatically every day, certain days of the week or days of the month, or variations such as every first and third Tuesday of every other month. As mentioned, a summer schedule could start on June 1 and end at the end of August.

Schedules can be customized to only execute under certain conditions. For example, a ceiling fan in the office could be scheduled to turn on to medium speed every week day, but only if home_status = "home", and the room temperature is above 70 degrees. A garage door opener could be scheduled to close every day at sunset, but only if it is open. A user might schedule the Amazon Grill and Chill playlist to play at cocktail time on weekend days. The possibilities for smart home schedules are endless.

Triggers

Another way to make home automation tasks happen is with the use of triggers. The simplest example is a button push. You push a button on a smart switch or a smart keypad and your hub receives the signal and makes something happen. The something could be as simple as turning a light on or off or as comprehensive as executing a Good Morning action group containing a list of numerous actions. The Good Morning trigger could be initiated by pressing a button on the MiniMote on the night table. The trigger would then cause the Good Morning action group to commence. See figure 18 for how you could do this in Indigo.

Triggers can be quite sophisticated. A water leak on the floor can trigger an action group which includes shutting off the water main to minimize damage and then sending an email to you or your landlord. Walking in your front door can be detected by a motion sensor and result in triggering which changes the variable home_status from "away" to "home". The reverse can also be useful. A trigger based on leaving home could change the variable home_status from "home" to "away" could cause the water main to turn off.

A trigger could be based on a numeric change in a variable. For example, if the sunroom temperature rises above 73 degrees, turn the sunroom ceiling fan on to medium speed. A trigger could be created based on sending an email to your hub. If the hub receives an email with "Status" in the subject line, the trigger could execute an action which compiles a list of variables and sends you an email with details on the status of any or all of the devices in your smart home.

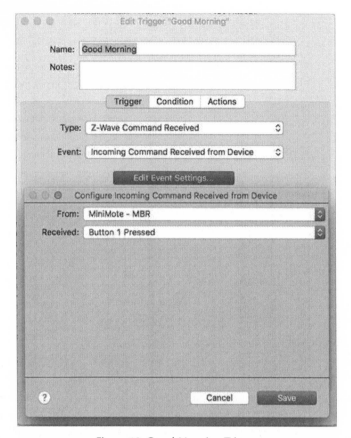

Figure 18. Good Morning Trigger

MultiSensor 6

A very useful device for creating triggers is the Aeotec MultiSensor 6. See fig-ure 19. The MultiSensor 6 has six sensors built into it. The following description will highlight an example or two of how each sensing capability can be used for a trigger.

1. The motion sensor has a range of more than 15 feet and a 120° field of view. Detecting motion can enhance your security system, and generate alerts. Monitoring presence can have other uses. If a child is somewhere you don't want them to be, you can be alerted. A more benign use is to have lights turn on whenever you enter a room.

Figure 19. MultiSensor 6. By Aeotec

2. The temperature sensor is capable of accurately measuring between 14° F and 122° F. Thermostats do their best to maintain a certain temperature, but the HVAC system cannot always meet the needs of certain rooms. I have a sunroom with a lot of windows which offers a beautiful view, but on really hot days, the air conditioning cannot keep it cool. A trigger is activated whenever the temperature exceeds 73 degrees. At this point, the ceiling fan turns on to medium speed. If, after a half-hour, the temperature if still above 73, the fan speeds up to high.

3. A digital light sensor can measure the level of light in a room or outdoors. The light sensor can help your smart home maintain its ambiance by setting lights to the perfect level, day or night. A trigger can be set up based on the level of light in a room combined with the time of day.

4. The humidity sensor can measure humidity ranging from 0% to 100%. Bathrooms and kitchens can get too humid, while a room finished extensively in wood may not get enough humidity to protect the wood. A smart home can have one or more triggers to automate when fans, motorized windows, dehumidifiers, and humidifiers go into action.

5. MultiSensor 6 contains a seismic sensor capable of detecting vibrations, tampering, or seismic activity within the home. Most clothes dryers make

an obnoxious sound when the dryer has finished its job. An alternative method is to create a trigger based on the presence or absence of vibration of your dryer. It can send you a text or email when the dryer stops running.

6. Finally, the MultiSensor 6 accurately measures ultraviolet light, the kind of light which can cause furniture fabrics to fade, or worse yet, eye damage. A trigger can cause solar shades to go down when the UV reaches a certain level.

IFTTT

In chapter 2, I described IFTTT, an abbreviation of "If This Then That". The IFTTT idea was developed for web services, but with the growth in home automation, new smart home ideas are emerging. A simple example is about weather. If wunderground.com has a forecast of severe weather approaching where you live, you could trigger not only to receive an email, but have your home audio system make an announcement and change the color of the room. Another example uses Geofencing. If home_status = "away" and you arrive to within 25 miles, a trigger could turn up the heat or turn down the air conditioning set point.

Most triggers cause something to happen when something else happens. If a button push, then turn on the music, if the door opens, turn on the light. If this, then that. However, there is another possibility. You may want something to happen if something else does not happen. This was the case for my backup generator.

The Generator

In chapter 5, I described the automatic weekly test cycle of the backup generator at our vacation cottage. On the day and time you select, by simply pressing the start button, the generator starts and runs for 15 minutes and then shuts down. From this point on, it will do the same thing every week at this same time. The weekly exercise of the engine prevents its oil seal from drying out and damaging the generator. It also provides some charging of the starter battery.

The weekly engine exercise works well, if the battery can start the engine. However, as I described in chapter 5, if I am not at the cottage, I have no way of knowing if the weekly exercise occurred. I have had several occasions over the past five years where the exercise did not occur because of a dead battery, a power failure occurred, and the generator did not do its job. In one case, I got a freeze warning message from my home automation system and, fortunately, my neighbor was at home. He checked the house and the power had been restored. A contractor came later and replaced the worn-out battery.

I thought there must be a way to monitor the generator and be sure it was able to make its weekly run, other than replacing the generator with a new one with a remote warning system. A replacement would be very difficult and expensive. I thought there must be a way some home attitude could be applied to the problem. I concluded I could combine a schedule, two triggers, and some actions which could solve the problem.

I tried several experiments with the home automation system and developed a solution. The first step was to purchase a Vision Z-Wave Shock, Vibration, and Glass Break Sensor, and attach it to the top lid of the generator using Velcro®. See figure 20. I set the generator to make the weekly test on Saturday morning. The next step was a script written using the Python scripting language, but like the Good Morning script, I will show the list of actions in plain language to make it clearer.

Figure 20. Vision Vibration Sensor. Photo by John Patrick

✓ The weekly generator test runs (or not).

✓ If it runs, the Z-wave vibration sensor sends a signal to the hub which causes a trigger.

✓ The trigger causes two actions: the variable gen_start is set to true, and the variable generator_start_time is set to the time the trigger occurred.

✓ When the generator stops running after the 15-minute test, it causes a second trigger which causes two actions: the variable gen_stop is set to true, and the variable generator_stop_time is set to the time the sensor no longer sensed the vibration.

✓ At 1 PM, a schedule runs and causes one of two emails to be created and sent to me.

✓ The emails use the variables described above.

✓ If the generator ran, the email says, "The Generator ran at 11:30 AM as planned. No worries."

✓ If the generator did not run, the email says, "The Generator did NOT RUN. Action required."

✓ The final step is to set the variable gen_start to false.

Summary

Schedules and triggers are the ingredients which can make your home smart. You can keep it simple or let your home attitude run. There is no limit to what you can do. An increasingly important trigger is your voice which is one of the user interfaces I will discuss in the next chapter.

CHAPTER 13
User Interfaces

In the not too distant future, we will be having conversations with our smart home, and it will do as we ask. Perhaps someday, our homes will be so smart they will manage themselves without our participation. Like autonomous self-driving cars, our homes may do everything they need to do to keep us safe and comfortable. However, in the meantime, hubs, schedules, action groups, and scenes can be most effective if there is an easy to use way for us to interact with them to make our home smart.

Homeowners interact when they issue commands directly by pushing buttons on a device such as a thermostat, door lock, light switch, or keypad. However, to initiate an action group, scene, trigger, or schedule, the homeowner must interact with a hub or instruct the hub to act on its own under certain conditions or schedule. The interfaces available depend on what kind of home automation hub you have, but most have at least a web page and a mobile app. In both cases the interface presents a menu or buttons enabling the user to select actions such as Good Morning, Lights Out, Open Garage Door, etc. Some hubs can accept an email as another way to execute actions. Probably the fastest growing method of interacting with a home automation hub is using our voices. "Alexa, turn on the kitchen lights" will become a very natural way to interact with our smart homes.

There are many user interfaces available to interact with your smart home. If I explained even a fraction of the possibilities, it would make this book too long to write and too long to read. I will select some examples I have found useful in my smart home.

MiniMote

As mentioned earlier, one of my favorite home automation user interfaces is the MiniMote. See figure 12. Eventually, the voice-based devices may be the ultimate, but for now, they are limited. The MiniMote is a compact remote with four buttons, and the buttons do exactly what you want them to do. Each button accepts two interactions: press and long press (press and hold for one second). These interactions give you eight possible triggers. You can place the MiniMote wherever is most convenient. There is no app to install, the MiniMote works directly with your hub. A charge by USB is only needed a couple of time per year. I have a handful of MiniMotes around the house. You might want to have one in every room. See table 11 below for what one of mine does.

Table 11. Night Table MiniMote Button Functions

Button	Function
Button 1 Pressed	Execute Good Morning Action Group
Button 2 Pressed	Execute Good Night Action Group
Button 3 Pressed	Toggle Night Table Lamp On/Off
Button 4 Pressed	Turn off lights and music. Light path to downstairs
Button 1 Held down	Increase volume of Sonos speaker
Button 3 Held down	Decrease volume of Sonos speaker
Button 2 Held down	Turn on ceiling fan to medium speed
Button 4 Held down	Turn off ceiling fan and fan light

Smartphone

Other than schedules and triggers, and the ultimate world of voice-activation, the most active user interface to our smart homes is the smartphone. Let me put "smart" in perspective. In 1976, Seymour Cray introduced the first supercomputer, the Cray-1. The term supercomputer meant it was the most powerful computer at the time. As for the Cray-1, it was super in many respects. It cost $5-$10 million, weighed more than 5 tons, and used as much electricity as ten homes. Super as it was, the Cray-1 had no app store, could not play a song, or even make a phone call. Scientists and researchers embraced the Cray-1 because it enabled them to perform scientific simulations and explore data at a speed not previously possible.

Fast-forward 41 years from the introduction of the Cray-1 to Apple's iPhone 8 and Samsung's Galaxy Note8. These devices are more than 100 times more powerful than the Cray-1 in every respect, and hundreds of millions of people around the world will be carrying them in their pocket or purse. The Apple and Samsung smartphones can determine if you are moving, how fast you are moving, your latitude and longitude, direction of travel, your pace, and the barometric pressure to determine your altitude. The smartphones and tablets we take for granted are truly supercomputers. With a connection to Wi-Fi or cellular Internet, our smartphones can act as a hub or communicate with a hub. They can turn devices on or off or execute actions while at home or remotely.

Some hubs can accept an email, so we can use our smartphones to connect to our smart home to ask for some information. As mentioned earlier, a homeowner could send an email to myhub@mysmarthouse.com with Security Status in the subject line. The hub would reply with an email listing the open or closed status of all the doors and windows of your home.

Control Pages

In addition to a full-function app on the Mac, Indigo Domotics offers the free Indigo Touch app for the iPhone. Indigo offers a feature called Control Pages. Control Pages are graphical pages you can create using Indigo which enhance the ease of use to control Indigo in a way which is unique to your smart home. You can easily customize the page design from the background image and color to the placement of text labels, control images, and what actions clicking on those images performs.

You can create one control page or as many as you want. I have one with a menu of rooms in the house. See figures 21 and 22 for examples of Indigo control pages on an iPhone. You can use an extensive menu of images included in Indigo. For example, you can use images of various kinds of lights and other devices which change color to indicate the status of the device. Any item on a control page can be linked to an action in the Indigo hub. Control pages are accessible either through a web browser or via the Indigo Touch app on an iPhone or iPad. You can create a comprehensive control center page showing the status of every device, the current value of variables, such as indoor and outdoor temperature, and a log of every action which has taken place in your smart home.

Figure 21. Indigo Control Page. By John Patrick (1)

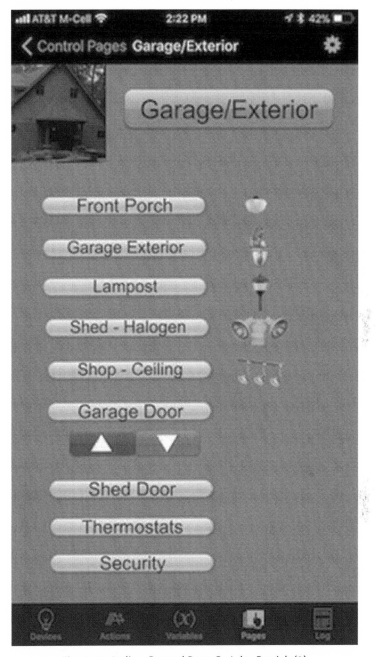

Figure 22. Indigo Control Page. By John Patrick (2)

Voice

The most natural way to communicate with your smart home is by voice. Talking to your home, sometimes called voice activation, and asking to turn on a light or put down the shades is not a new concept. In the 1968 science-fiction film, *2001: A Space Odyssey*, Dr. Dave Bowman, played by Keir Dullea, used voice activation with the space station's intelligent onboard computer, H.A.L. 9000. "Open the Pod bay doors, HAL", he said. After no response, he said "Hello, HAL, do you read me?" After a few requests, HAL said, "Affirmative, Dave. I read you." After an unsuccessful dialog, HAL wrapped it up with, "Dave, this conversation can serve no purpose anymore. Goodbye."

I tried voice activation in my home some years ago. It was not as threatening as HAL taking over the space station, but I found it to be very unreliable. The microphone and translation technology was primitive. I would find myself yelling at my home and often getting no action or, worse yet, the wrong action.

Today, controlling things by voice has become very sophisticated with the advent of highly accurate voice recognition and the use of artificial intelligence (AI) to understand your words and interpret what you want done. As of this writing, Amazon has a significant lead over Apple and others with its Echo technology. If I say, "Alexa, turn on the office lights", it works 100% of the time. Echo does not interface with all hubs, and not with all devices on hubs it does work with. At this stage, Echo works with certain devices and can turn them on or off or set a thermostat set point. It does not yet have the ability to execute an action group. I am sure it will.

I am confident all the voice activation systems will work with most hubs and devices in the near future. The breadth and depth of things you can do with voice will continue to expand. When you say, "Alexa, let's party", the Echo will communicate with the hub, determine what room you are in by motion sensing, look at the calendar to see how many people are coming, lower the heat set point on the smart thermostat to compensate, look at what music you played and the lighting colors and levels from the last party, and get everything ready.

As with all technology, there can be a dark side. Andrew Liptak wrote in *The Verge* about a six-year-old in Dallas, Texas who was talking with her family's new Amazon Echo. She said, "Can you play dollhouse with me and get me a dollhouse?" The Echo readily complied and ordered a $250 KidKraft Sparkle mansion dollhouse.[109] The parents figured out what happened and updated their voice purchase settings in the Amazon Alexa app. They could have returned the product, but instead they donated the dollhouse to a local children's hospital.

Our voice will play a major role as we get comfortable with a home attitude. All the major players are making large investments to make voice a natural way to interact with their technology, products, and services. The current leading voices are Amazon's Alexa, Apple's Siri, Google Home, Microsoft's Cortana, and Samsung's Bixby.

Polly

Another kind of voice technology is text-to-speech. Amazon Web Services offers a free (within limits) text-to-speech service called Polly. Polly can turn text into lifelike speech, allowing you to create applications which talk to you. Polly uses advanced deep learning technologies to synthesize speech which sounds like a human voice. The service includes dozens of lifelike voices in a variety of languages. Polly is what I used for the Good Morning action described earlier.

Artificial Intelligence

Of all the things which are candidates for the "next big thing", I have no doubts about what will be the biggest: artificial intelligence (AI). The term AI has been around for more than 50 years. It has flared up in interest as a "big thing" on multiple occasions and then flared down, but this time, it is for real, in my opinion. The reason I say this is because of two exponentially growing factors. First is the amount of data available about just about everything: data from sensors, tweets, emails, postings, GPS coordinates, medical metrics, invoices, micro-payments, Internet searches, and home automation requests. In the past, it was too costly to store everything. Today, storage is extremely inexpensive and makes it possible to save everything. Consumers saving tens of thousands of pictures on multiple devices is a tiny example. The growth is exponential.

The other factor is the processing power of computers. As mentioned earlier, an iPhone is more than 100 times as powerful as the first supercomputer of 1976. Google, Amazon, Microsoft, IBM, and others are building web servers with even more incredible power. Estimates vary, but there are probably 50 million servers currently operating. The growth is exponential. When something grows exponentially, all of a sudden, the growth takes off. See figure 23 for a graphical concept developed by Tim Urban. [110]

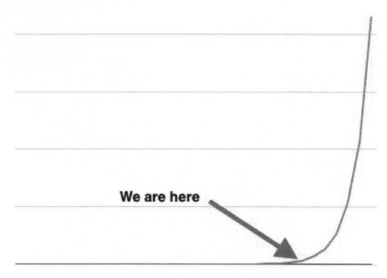

Figure 23. Exponential Growth as described by Tim Urban

Tim Urban is the author of *Wait But Why*. He said that the arrow shows where we are with AI.[111] I agree. The combination of vast amounts of data and a vast amount of computing capacity to make sense of the data will make the computers smarter than humans and just about every aspect of our lives will be changed as a result. This includes home automation.

We will no longer need to say "Alexa, I am cold". The AI behind your voice activation system will know every request you make, when you make it, and eventually, the AI will know why you made the request. Based on the accumulated knowledge, the AI will anticipate, it will know the temperature, and know what you will consider to be cold. It will change the thermostat setting, and you will not get cold.

Summary.

User interfaces will continue to become more powerful and intuitive. Smartphones will get smarter and smarter. AI will become part of nearly all technology. The obvious question is whether we want the AI to know all about us? It is much more than whether we are cold or not. Privacy, as discussed in chapter 8, will become an increasingly important topic regarding home automation. The next step is to determine how you will convert your home attitude into an action plan.

CHAPTER 14

Home Automation: How to Get It

At this stage, I hope you have developed a home attitude and are ready to learn how to make your home smart. There is a wide range of possible deployments. Many years ago, I developed a mantra for how to approach new opportunities such as home automation, the Internet, or almost any hobby. In *Net Attitude: What It Is, How to Get It, and Why Your Company Can't Survive Without It*, I described the mantra as Think Big, Act Bold, Start Simple, Iterate Fast.[112]

Thinking big means to envision the many possibilities of how a home attitude can enable you to have a smart home which enhances your safety, security, energy savings, entertainment, and convenience. Act bold means to make a purchase of some home automation technology and get your feet wet. Most important is to start simple. For example, you might purchase a Wemo® Switch Smart Plug, made by Belkin.[113] See figure 23. Simply plug the Wemo® into a wall receptacle and then plug a table lamp into the Wemo®. Turn the lamp on and then press the button on the Wemo®. The button lets you turn the light on and off. Next, download the Wemo® smartphone app, and then you can experience turning the light on and off from your phone. Setup a schedule with the app so the light comes on automatically at sunset and then goes off at 11:00 PM. You can also connect your Wemo® with an Amazon Echo and ask Alexa to turn the light on or off for you. After these baby steps, you will see your home attitude coming to life.

Figure 24. Wemo® Switch Smart Plug. By Belkin

The last part of the mantra is to iterate fast. This means continuously adding devices and capabilities to make your home smarter and smarter. I started with just one light switch back in the late 1980s. Today, I have 127 devices, 111 triggers which cause the devices to do various things, 18 schedules which turn things on and off, and 117 action groups which enable my home to do many smart things.

The sections to follow dive deeper into the three approaches to implementing a home automation system to make your home smart. The first is to go professional and hire a systems integrator. Second is to subscribe to a home automation service from a communications or security services provider. The third approach is DIY. At the end of the chapter, I will summarize the pros and cons of each approach.

Go Professional

My early forays into home automation began in the late 1980s. I bought some devices at Radio Shack and setup a few lights to go on and off automatically based on the daily sunrise and sunset. What I did was elementary, but it gave me a glimpse of what was possible. During 2001, while planning a new home, I decided to add as much home attitude is possible. At the time, there were not many devices available, but I discovered a security company which had strong expertise and an innovative sales engineer.

The sales engineer and I met weekly for nearly a year to design the various capabilities I wanted. Some were off the shelf, but some we invented from scratch. For example, I wanted an automated garage door opening system. I did research to find equipment which could do what I wanted, and the sales engineer figured out how to install it and make it work reliably. The solution involved a small pizza box sized antenna mounted on an 18-inch-high post near the entranceway to the garage doors. Each car got a bar coded sticker for the car window. When a car passed by the pizza box antenna, the home automation hub would recognize the trigger, know which car it was, and open the appropriate door automatically. Upon leaving the driveway, the opened door would close. This solution is clearly in the convenience category, but I learned a lot from it for future ideas.

Working with an integrator resulted in a successful implementation of home attitude, but I was embarrassed when BusinessWeek Magazine wrote a story about the home titled, "King of His Digital Castle".[114] I was proud of the house, but did not consider it a castle. I now live in a different smart house, but the unique feature of the 2002 smart house was the high degree of integration. The integrator said they had done many homes with automated lighting, many with extensive audio/video capabilities, some with numerous sensors and triggers, but none which tied all those things together in a single home automation hub.

Professionally built custom systems are available from companies which operate as systems integrators. They typically are business partners of companies such as Control4, Crestron, Leviton, Lutron, and Savant all of which make home automation products. The integrators either offer security monitoring or partner with a company which provides it. When searching for an integrator, you will find plenty to choose from. Many local security companies have evolved into the home automation industry. There is a parallel to installing door and window sensors and hooking them up to a security panel. It is like connecting home automation devices to a hub. Home audio video companies have similar skills. Network support companies which provide local area networks in homes are also natural candidates. The key thing with any of the companies is to test if they have the vision to implement home attitude to the level you want. Do not hire just anyone who says they can install devices.

New companies and breakouts from existing companies are forming to go after the home automation industry. Pune, India based MarketsandMarkets is a large market research firm serving 1,700 global enterprises with consulting studies. They valued the global home automation market at $39.93 Billion for 2016, and expect it to grow more than 11% per year over the next five years.[115] The following paragraphs describe some of the leading home automation companies offering professional installation of their systems.

Control4

One company which is gaining a lot of momentum in the home automation market is Salt Lake City, Utah based Control4 Corporation, founded in 2003. Control4 provides home automation solutions for homes and businesses in the United States and internationally. As of May 2017, the publicly traded company was valued at $450 million.[116]

Control4 offers a comprehensive product line ranging from a single room light controller to a full-blown home automation system touching every aspect of the largest of homes. The Control4 approach to creating a smart home is to provide a software system which ties together audio, video, lighting, temperature, security, communications, and other home automation devices. The software runs on dedicated special purpose home automation hubs called Controllers. Control4 sells its products through direct dealers and distributors.

Creston

Crestron Electronics, Inc. was founded in 1971, and is based in Rockleigh, New Jersey.[117] The company designs, engineers, manufactures, and supplies home automation solutions for consumers and businesses. It offers multi-room audio and video solutions, including speakers and audio distribution equipment. In addition, it offers solutions for lighting and shading. Crestron also makes touch screens, remote controls, keypads, and application software. It serves residential, corporate, education, hospitality, government, transportation, resources, and retail markets.

Leviton

Leviton, founded in 1906, family owned, and based in Melville, New York, is possibly the oldest home automation company in the world.[118] The company offers everything from simple switches and receptacles, to networking systems and smart home automation, LED lighting and lighting energy management systems, and security solutions.

Lutron

A young physicist in New York City named Joel Spira had a radical idea in the late 1950s. He was fascinated by the aesthetic manipulation of light, and invented a solid-state device which could vary the intensity of the lights in a home. It was a radical idea at the time because lighting control was complicated, expensive, and required bulky rheostats which used a lot of energy and generated a lot of heat. Lighting controls were used primarily to dim stage lights in theaters. Consumers would never think of having dimmers in their homes, until Spira created a solid-state dimmer which could replace the light switch in a standard residential wall box.

Today, based in Coopersburg, Pennsylvania, the company, which Spira named Lutron Electronics, offers a broad portfolio of services for design, implementation, and support of all forms of lighting.[119] Lutron holds more than 2,700 worldwide patents including hundreds of lighting control devices and systems. The company offers more than 15,000 products. Lutron has also innovated window shade technology for the control of daylight, as well as wired and wireless systems to integrate the control of both daylight and electric light. Lutron lighting controls are installed everywhere from single-room apartments to palatial homes, including the White House and Windsor Castle.

Savant

Savant, based in Hyannis, Massachusetts, was founded in 2005. The company has focused on ease of use and claims to offer the best experience in home automation. Savant's website makes it clear the company caters to luxury homes.[120] The company site says, "We've become the home automation brand of choice for the world's most luxurious homes, castles, and even yachts."

Savant claims to be the first home automation company to embrace mobile technology. The user interface and experience is very similar to Apple, but they also support Android. Numerous mobile apps are available to provide a complete set of home automation control features.

Unlike most other home automation companies which offer proprietary touchscreens, Savant offers a range of iPad wall and table mounts. For many consumers, this means having a familiar interface across all the various home automation tasks. The Savant system provides energy monitoring tools so you can see how much electricity your home consumes. A feature unique to Savant is a universal remote you can set up on your own to control home theater and whole-house audio equipment. The remote is an excellent way to experiment with various home automation capabilities before you spend the money to have a professional installation across your entire home.

Go Professional Summary

The five companies I have highlighted all have excellent products and services. They work with dealers, distributors, and installers to provide professional design and installation of home automation solutions. Choosing a professional approach should lead to you getting exactly what you want, and have a reliable system which can serve you for a long time.

One thing the professional integrators have in common is pricing. All the reviews I looked at extolled the various advantages of each company, while the disadvantage of each was noted with the "expensive" label. A simple installation of a smart lightbulb may cost $100 or less. A "whole home" system can cost $1,000 per room or more.

Another disadvantage is the lack of flexibility to make small changes to how your smart home works. You may want to simply add another receptacle, change how an alert works, or add a new function to your Good Morning scene. Generally speaking, it would be necessary to bring the installer back to make the changes and incur the expense of a service call or programming changes. The integrator who installed my first smart home charged $150 per hour for any programming charges. It can add up quickly.

For those who want the benefits of home automation without having to know too much about how it works, then the professional approach can well be worth the investment. Knowing there are experienced installers on call when needed can provide peace of mind and allow you to effortlessly enjoy living in a smart home.

Subscription Based Services

Subscription based smart home services are typically offered by telecommunications providers such as AT&T, Comcast, and security monitoring companies such as ADT, Safe Streets USA, and Frontpoint. These companies are already providing phone or security services for the home, and it is natural for them to extend their services to include home automation. It is a natural extension for them, but you should consider if it is a natural extension for you.

ADT

ADT Security Services has been around for more than 100 years, has 20,000 employees, and had nearly $1 billion in revenue for 2016. Their focus is security, but they also offer home automation for a monthly fee. The fee is typically $50 including the security monitoring. The home automation features include most of the basics.

You can arm your security system remotely using the ADT Pulse® smartphone app, and receive a text message or email when your primary entrance is opened. The system includes remote temperature control which ensures your smart home is the temperature you want when you arrive, and you can set the system to alert you if your home gets too warm or cold.

You can turn lights on and off from an ADT keypad, from the app, or based on a trigger, such as a door opening. If you have smart door locks, you can lock or unlock them from the keypad or your smartphone app. Other custom alerts can turn on a light if your motion detector is triggered, or disarm the system and turn off the thermostat if your smoke detector is triggered. You can also setup schedules for when a light or other device goes on or off.

Safe Streets USA

Safe Streets USA is one of ADT's oldest and largest authorized dealers. As an ADT Authorized Dealer, Safe Streets USA only sells and installs ADT-monitored security systems. The company has a network of nearly 300 trained installation consultants operating in 44 states. They provide same day installation of your system, and train you on how to use your system and all its features.

AT&T

AT&T offers 24/7 monitoring with remote access from your smartphone, tablet, and desktop or laptop computer. They provide text and email alerts. Communication from your home to their system is wireless, no landline required. In addition to live or recorded video of your home, you can remotely lock or unlock your door, manage temperature, and small appliances. Devices they support include water leak detection and temperature sensors, keypad, keyless door locks, indoor siren, door and window sensors, motion, smoke, glass break, and carbon monoxide sensors.

Comcast

Comcast has an offering called Xfinity Home Secure which costs $24.99 per month under a two-year contract. Their website describes the offering as 24/7 security and professional monitoring with home control and live video monitoring for an additional $9.95 per month. The basic security package includes 3 door or window sensors, a motion sensor, touchscreen controller, and a wireless keypad.

For other home automation devices, Xfinity partners with smart home brands such as GE, Lutron, and Philips lighting, Kwikset door locks, Chamberlain garage door openers, and Google's Nest Learning Thermostats. The website is light on details, but Xfinity offers a personal assessment of your home and current equipment, an evaluation and customized security system recommendation, and a tutorial on how to use their system.

Frontpoint

ASecureLife.com, started in 2008, provides research to inform consumers with up-to-date and transparent information about security companies, including pricing, contract lengths, cancellation policies, and customer service history. They selected Frontpoint Security as the best home security system for 2017 with a 4.0 out of 5.0 in their annual best home security systems review.[121]

The monitoring service does not require a landline. All of their plans provide a dedicated cellular service which Frontpoint uses whenever your in-home security system needs to communicate with the monitoring central station. Redundant central monitoring stations provide increased reliability. The review gave high marks for excellent customer service, interactive monitoring via your smartphone, and a 30-day money-back guarantee. The monthly fee ranges from $34.99 to $44.99 depending on the options you select, and they require a 36-month contract.

Frontpoint supports numerous home automation devices, but they do not provide installation service. Their offering is a hybrid of professional monitoring with do-it-yourself device installation. Although the monitoring and customer support get high marks, a PC Magazine review described their offering has having, "Pricey monitoring plans and accessories."[122]

Subscription Service Summary

The subscription services offer many conveniences. There is a clear benefit and peace of mind resulting from their direct contact with police, fire, and medical responders. However, when it comes to home automation, some people may have concerns about the privacy of their home data and prefer to keep their lighting, music, thermostat settings, and other data about what is going on in their home, inside their home. This is my personal view as well.

The largest consideration, in my opinion, is the degree of customization provided by subscription based services. They all offer the basic capabilities to turn lights on and off, set up schedules and triggers, and receive alerts when certain things happen. However, if you want to push a button on a remote and have a series of Good Morning actions take place including lighting scenes, music selections, announcement of the weather forecast, adjust the thermostats, raise the solar shades, and turn on the irrigation system, it is unlikely the subscription based services can handle the task. In my opinion, sophisticated and unlimited home attitude is best achieved with a professionally designed or do-it-yourself system.

Do It Yourself

The third choice for implementing your home attitude vision is Do It Yourself (DIY). A DIY approach can include all or just a part of the smart home. Although there are DIY security monitoring solutions, most people choose to use one of the professional monitoring services for this critical aspect of the smart home. If something breaks with your automated lighting or music, the consequences are small. If the window and door sensors are not working or the automatic call to police, fire, or medical services fails, the consequences can be life threatening. A DIY security solution can work, but when it comes time to make the call to one of the emergency services, emotions and other factors can take over and inhibit you from doing the right thing.

If you are comfortable with taking responsibility for your security system, there are many choices. For example, the SimpliSafe Home Security System provides affordable hardware and reasonable monthly monitoring fees. The wireless window and door sensors are easy to install with no wiring or drilling required. SimpliSafe includes cellular connectivity and battery backup.

Similar offerings are available. The Smanos W020i Wi-Fi Alarm is a DIY security system which is easy to install and is expandable, but it doesn't integrate with other smart home devices. The SkylinkNet SK-200 and the iSmartAlarm Premium systems are great for homeowners with a small budget who want to save money by monitoring their homes themselves. The drawback to such systems is they leave it up to you to act when an alarm is triggered. They don't check in to see if everything is OK when your smoke alarm goes off or if your front door alarm trips, and they don't call the local authorities.

Once your security solution is in place and tested, you are ready to consider applying home attitude to make your home as smart as you want. The security system can remain independent of your home automation system, but you may want to connect them later. The ultimate smart home is one where every sensor, including the doors and windows of the security system, plus your lighting control, digital door locks, smart thermostats, irrigation system, home audio and video systems, and even your smart flower pot on the deck are all integrated into one system which you can control from your smartphone.

The biggest challenge in DIY is the same as the biggest advantage, unlimited possibilities. Before you can do it yourself, you must know what it is you want to do. Like all consumer electronic devices, home automation devices have increased in capability and decreased in price over recent years. The trend is continuing and a plethora of devices are becoming available. Sensors, switches, keypads, remotes, smart thermostats, digital door locks, smart ceiling fan controls, and high-fidelity speakers have become more affordable and readily available online. More and more home automation devices are wireless, making them much easier to install than having to run wires and cables throughout the home.

Remember the mantra, Think Big, Act Bold, Start Simple, Iterate Fast as discussed at the beginning of the chapter. DIY fits this philosophy very nicely and offers quite a few advantages. You can proceed at your own pace. You are not buying a package deal, you are creating your own package and deploying it on your own schedule. You don't need a contract. Perhaps most important, you can mix and match the various devices which you feel offer the best value for you, not the ones which are made or sold by the installer or service provider. You can comparison-shop and purchase devices directly from a manufacturer, from a general online retailer, or from a retailer which specializes in home automation products.

Devices such as switches, receptacles, thermostats, and door locks can work independent of your hub. In other words, they can be dumb devices, but when you are ready, you can connect them wirelessly to the hub so they become smart. A smart switch will turn on and off with the press of a finger, with no assistance from the hub. Once you connect the switch to the hub, it can then be turned on and off by commands from the hub. You can have it both ways, manual and automated. This makes the DIY approach easy to adapt to the Think Big, Act Bold, Start Simple, Iterate Fast mantra.

The best place to start is to select what hub you will use. Every smart home must have a hub. A hub can be a specialized device or it can be a desktop, laptop, smartphone, or tablet which has specialized home automation hub software. As discussed earlier, hubs act as a central controller for all the devices which are part of your home automation system. It is the hub which enables a home to become a smart home.

In the following paragraphs, I will describe companies which specialize in home automation solutions for DIYers. Each has one or more hubs to choose from plus a wide range of home automation devices. They all have good support resources or active communities of home automation hobbyists who share their knowledge and experience with others.

Apple

Apple has made it clear it wants to compete in the emerging home automation market. The hub is an app on an iPhone or iPad called Home. In classic Apple style, Home offers a clean and easy to use interface. Apple provides home automation device makers with specifications the company calls HomeKit. Products adhering to the HomeKit specs will work with the Home app. HomeKit was announced in June 2014 and the Home app was announced two years later. It was a slow start, but device makers are adopting the HomeKit standard and dozens of devices now work with Home. Apple CEO, Tim Cook, promised hundreds of new HomeKit-compatible devices by early 2017, and introductions of newly supported products are underway.[123]

The Home app supports scenes, schedules, automated actions, delegation to family members or other users, and many other special features. For example, Geofencing allows you to activate a scene or device when you enter or leave an area. For voice assistance, Siri can turn lights on or off, but is expected to have additional capabilities as developers get experience with HomeKit.

In June 2017, Apple introduced the HomePod home audio speaker. In addition to an impressive speaker design, the HomePod has six microphones which allow listeners to talk to Siri and select Apple Music, ask questions about nearly anything, and control HomeKit devices.

From a DIY perspective, you can do-it-yourself with the Home app if you select HomeKit devices. If you want to include devices which are not HomeKit devices, then you cannot do it yourself. This is the two-edged sword of Apple. There are thousands of devices out there. If you already have some of them which are not HomeKit devices, then you cannot build a fully integrated system. Time will tell whether all the device makers will adopt HomeKit. If they do, Apple stands to be a major, possibly dominant, player in home automation.

e-WeLink

e-WeLink is a free smart home app from China which can manage and control more than 100 smart devices from 80 brands. e-WeLink supports Apple HomeKit, Google Nest, Amazon Echo, and a number of Chinese brands. e-WeLink supports multiple languages, and the website says the number of supported languages will increase in the future. e-WeLink is relatively new, but is getting a lot of attention, with over one million active users. The platform hub runs on Apple and Android smartphones.

Another Chinese home automation thrust is coming from ITEAD, based in Shenzhen, China's biggest electronics market. It is part of a highly integrated global electronic supply chain. ITEAD targets DIYers with high-quality, innovative products at very low prices. Products made by ITEAD are supported by the e-WeLink platform.

ITEAD and e-WeLink, along with retailer giant AliExpress are trying to fuel an innovation revolution in home automation with easy to use prototyping modules, low-cost development platforms, and custom-made solutions. I expect what these companies are doing will be of great interest to curious students, engineers, and hobbyists who love to create.

HomeSeer

Bedford, New Hampshire based HomeSeer has been a manufacturer and provider of home automation systems since 1999. [124] HomeSeer has been a strong home automation industry player with many positive reviews and thousands of systems installed worldwide. HomeSeer has partnerships with many home automation industry giants, although their partnership with Apple did not go well, and HomeSeer is not part of the HomeKit ecosystem.

HomeSeer calls their hubs Home Trollers. The hubs come in five different models ranging from $200 to $1,000 depending on the features and capacity you choose. The Home Trollers work with thousands of products from hundreds of manufacturers using most popular home automation technologies. You can also purchase HomeSeer software and create your own hub on a PC which you can dedicate to home automation.

Support for HomeSeer products is available from dealers or from their extensive online community forum. Do-it-yourselfers tend to be very good at sharing. In many ways, it is a better approach than relying on the experience of a dealer or even one of the major professional home automation companies. For any problem or question you may have, you can be sure someone in the community has an answer or a solution.

Indigo Domotics

Perceptive Automation, LLC is a Dallas, Texas based home automation company operating under the name Indigo Domotics. The company was founded in 2002 by Matt Bendiksen, formerly Director of Engineering at Macromedia,

Inc. In 2008, Jay Martin, formerly Vice President of Engineering at CyberSource Corporation joined Indigo Domotics. The two highly technical professionals have more than 50 years of engineering experience.

The company is small compared to some of the other DIY companies, but the owners have built an advanced DIY smart home platform called Indigo for the Apple Mac platform. The Indigo software is very easy to use, but also has highly advanced capabilities for those who need it. Excellent support is provided through a How-To-Wiki, User Forums, and an extensive library of useful tools and files. Indigo's biggest strength is the extensive community of users who not only help each other out on the user forum, but have also created some very high-quality plugins which extend Indigo's usefulness by supporting hundreds of additional devices and protocols.

I have been using Indigo since 2012 and currently use it to manage more than 200 home automation devices at my primary and vacation homes. The most important feature of Indigo, in my opinion, is the way it supports new devices. Commonly used devices have support built in to Indigo, but if something new comes along which Indigo does not support, you can likely find plugins, scripts, and applications in the User Contribution Library which expand the functionality in Indigo. For example, one plugin allows Indigo to work with Amazon Echo. Another Indigo plugin allows the Sonos home audio system to play an announcement of the weather forecast using an artificial intelligence (AI) capability from Amazon Web Services called Polly and another plugin for Weather Underground.

Anyone can create a plugin and share it with the entire Indigo Domotics user community. Plugin developers openly share what they have created and provide tips and techniques to others. If someone invents a new electronic mouse trap, he or she can create a plugin for Indigo which can enable the mouse trap to be supported as well as off the shelf standard devices. As of August 2017, the Indigo community has developed 213 plugins.

Smarthome

Smarthome, a SmartLabs, Inc. company based in Irvine, CA, is one of the world's largest home automation retail superstores. A comprehensive website and well-staffed product specialist telephone support provides an easy-to-use source for thousands of lighting, security, home entertainment, and other home automation products the average DIY enthusiast can safely install.

Smarthome distributes a proprietary line of home automation products branded INSTEON®. The products use a wireless home-control networking technology. Insteon devices are available for switches, keypads, ceiling fan controls, lighting control, door and motion sensing, and garage door control.

Smarthome has designed its Insteon hubs and devices to work very nicely together, but not necessarily with other vendors' hubs and software. Insteon devices are supported by Indigo, but not by SmartThings. They work with HomeSeer with the addition of a software plugin. Proprietary devices such as Insteon are reliable and easy to deploy, but can be limiting if you want to use devices from other vendors.

SmartThings

SmartThings, based in Mountain View, California, operates independently as a wholly owned subsidiary of Samsung Electronics. The Samsung SmartThings Hub connects wirelessly with a wide range of smart devices including lights, switches, outlets, sensors, cameras, doorbells, door locks, thermostats, and speakers. The hub allows the devices to work together, and you can monitor and control the connected devices using a SmartThings app for iPhone or Android. SmartThings also works with Amazon Echo or Google Home voice assistants.

The SmartThings Hub can send you alerts from connected devices when there's unexpected activity in your home, and can turn lights on or off when doors are opened or a person enters a room. Using what Samsung calls SmartThings Routines, you can set up a series of actions to take place corresponding to Good Morning, Goodbye, or Good Night routines. SmartThings supports many devices, but not all devices. There is no support for plugins.

Vera

Vera Control, Ltd. is a Hong Kong based home automation company. It was founded in 2008 by a group of entrepreneurs with backgrounds in wireless technology, consumer electronics, and software development. Vera hubs range from approximately $75 to $300 and can support more than 1,000 connected devices.

Vera hubs use the MiOS software platform. A strength of MiOS is the comprehensive interoperability with most major home automation products on the market, enabling Vera hubs to bridge any or all of them. Vera is web based and enables users to connect to it with any browser from any smartphone, desktop, laptop, or tablet.

A strength of the Vera solution is compatibility with many kinds of devices. However, the functions which the Vera can perform are somewhat limited for advanced DIY enthusiasts. Vera is easy to use and there is a strong community of users around the globe.

Do It Yourself Summary

Creating a DIY home automation solution requires some knowledge and skills, not as much as a few years ago, but nevertheless, non-trivial. You do not need to be a computer programmer, but you will need to learn how to create schedules, triggers, and actions which will be key to allow your home to be smart. Configuring your hub to do exactly what you want can be time consuming. One thing I learned many years ago is software does exactly what you tell it do, not what you want it to do. This can be frustrating at times. Many of the devices, such as a door sensor, are very easy to install with just basic tools, if any. Wall receptacles, switches, fan controllers, thermostats, and digital locks can be more difficult, but most any electrician or HVAC technician can do the job for you.

It is possible to mix and match from the three alternative paths, but I do not recommend this. The power of home automation is integration, tying everything together. Multiple parties in the same arena can lead to finger pointing in the event of problems. See table 12 for a summary of the Pros and Cons of Home Automation Alternatives for the three alternate paths to implementing your home attitude.

Next Steps

Now you have a home attitude, and are ready to make your home a smart one. Use the checklist below to help build an action plan to get started. Remember the mantra: Think Big, Act Bold, Start Simple, Iterate Fast.

✓ Make a list of the most important applications of home automation you would like to have such as lighting control, home audio, ceiling fans, door locks, etc.

Table 12. Pros and Cons of Home Automation Alternatives

Alternative	Pros	Cons
Go Professional	Get exactly what you want Benefit from experience of professionals Available service, upgrades, modifications	Can be expensive May not be flexible Cannot make tweaks yourself
Subscription Based Service	Less time consuming Standard capabilties may meet basic needs Predictable expense	Not as customized May not meet all needs Contract may lock you in
Do It Yourself	Potential to get exactly what you want Freedom to make upgrades and changes Can adopt newest technology at will Complete integraton of all capabilities Can be rewarding	Requires knowledge and skill Takes time to implement Takes time to refine Compatability can be limiting Can be frustrating

✓ Evaluate the pros and cons in table 12 and determine which alternative is the best for you.

✓ If you are leaning toward engaging with an integrator to obtain a custom solution, or toward signing up for a subscription based solution, begin an online search for those in your area. I would suggest evaluating at least two solutions so you can get a good comparison. Talk to some of their customers in the area where you live.

✓ If you choose to go the DIY route, visit the wikis or support forums of one or more of Apple, e-WeLink, HomeSeer, Indigo Domotics, Smarthome, SmartThings, and Vera or others. Search for things of interest such as door locks, thermostats, etc. and read about the experiences of the hobbyists and users.

✓ If you are going with DIY, select the hub which best meets your criteria in the first item in the list. Please note you may need more than one hub to meet all of your criteria.

✓ Purchase a few devices, install the hub, and start experimenting.

✓ Once you are comfortable with a few, lay out a plan to build out your smart home with additional capabilities.

✓ Maintain your home attitude, and evaluate new devices as they come to market.

✓ Always keep security and privacy at the top of your mind, especially when configuring new hubs or devices.

✓ Relax and enjoy your smart home.

Epilogue

More than 40 years ago, science fiction writer Ray Bradbury predicted homes would be interactive in the future. He envisioned homes would also be able to do things on their own, even after their human owners had died. The future has arrived. Skip Prichard, President and Chief Executive Officer of OCLC in Dublin, Ohio, said this about *Home Attitude*, "It's funny, but the title actually makes me think the home HAS an attitude – the AI makes it somewhat alive with thoughts."

AI will permeate every aspect of our lives. It is hard to predict exactly what the impact will be. Billions of dollars and thousands of engineers are at work building AIs which will be gathering and learning from all of our interactions with products from Amazon, Apple, Google, Microsoft, Samsung, and others. How much do we want them to know about our preferences, how much time we spend in each of our rooms, what music do we listen to, what time we go to sleep, what time we start our day, and much more? The era ahead of us offers a lot of new capabilities, but we should be cognizant of security and privacy as never before.

Acknowledgements

I am grateful for the time and interest of the following who contributed their ideas, quotations, or comments on early drafts: Bilal Athar, Mike Butler, Ken Ducey, Konrad Gulla, Ron Gruner, Jim Kollegger, Nick Lagaros, Eric Lutker, Harris Miller, Dan Ohlson, Aaron Patrick, Bob Patrick, Jason Patrick, Joanne Patrick, Tracy Patrick-Panchelli, and Skip Prichard.

Appendix A

Analog music is captured with recording equipment and then placed on the CD in CD-DA or digital audio format. This is done by electronically sampling the sound 44.1 thousand times per second and capturing two characters (bytes) of information about the characteristics of each second. This results in 88,200 bytes of data for one second of music. Multiply X2 for stereo and you have 176,400 bytes of data per second. Multiply times 60 and you get 10.584 megabytes per minute of music. A CD holds about 660 megabytes of data so giving you approximately 62 minutes of music on a CD.

OK, so what is MP3? There was a group of experts (from IBM and other companies) called the "moving pictures experts group" which created a standard called MPEG. MPEG has various "layers" which specify how audio or video can be compressed. Compression removes bits from the sampling process which are not essential or even recognized by the human ear. A brief pause in a song, for example, can be eliminated or compressed and then decompressed later when it is played. The result of compression is a much smaller amount of data which needs to be stored.

MPEG layer 3 (MP3) describes a particular standard for achieving high quality sound with compression. It results in a compression ratio of roughly eleven. In other words, with MP3 you can store roughly 11 hours of music on a CD. It also means CD music can be stored on a PC in about one eleventh of the space required if it were not compressed. It further means with compression it has become practical to send music over the Internet in a reasonable amount of time. The result was Napster, which was designed to share PC files, instantly became a very convenient way to share MP3 music files.

Notes

1   John R. Patrick, "Anyone Remember Heathkits?," *Attitude LLC* (2000), http://www.attitudellc.org/anyone-remember-heathkits/

2   Steve Hamm, "King of His Digital Castle," *BusinessWeek* (2007), http://www.bloomberg.com/bw/stories/2007-01-21/king-of-his-digital-castle

3   Ibid.

4   John R. Patrick, "In the News," *Attitude LLC* (2008), http://www.attitudellc.org/points_of_view/in-the-news-3/

5   Ray Bradbury, *There Will Come Soft Rains* (Old Greenwich, CT: Listening Library, 1976).

6   Ibid.

7   Daniela Hernandez, "Before the iPad, There Was the Honeywell Kitchen Computer," *Wired* (2012), http://www.wired.com/2012/11/kitchen-computer/

8   Ibid.

9   Martin E-Man, "X10," *BUILDYOURSMARTHOME* (2014), http://buildyoursmarthome.co/home-automation/protocols/x10/

10   Julie Jacobson, "Microsoft and Intel Promise Interoperability through New Home API," *CEPro* (1998), http://www.cepro.com/article/microsoft_and_intel_promise_interoperability_through_new_home_api

11   "A Brief History of Microsoft's Failed Attempts at Home Automation," *CEPro* (2014), http://www.cepro.com/article/print/tbt_a_brief_history_of_microsofts_failed_attempts_at_home_automation/

12   Joan Engebretson, "Understanding Home Automation Standards," *SDM* (2016), http://www.sdmmag.com/articles/92108-understanding-home-automation-standards

13   John R. Patrick, *Net Attitude: What It Is, How to Get It, and Why Your Company Can't Survive without It* (Cambridge, MA: Perseus Publishing, 2001).

14   "Total Number of Websites," *Internet Live Stats* (2015), http://www.internetlivestats.com/total-number-of-websites/

15   Omer Rachamim, "How Many Online Stores Are There in the World?," *Internet Retailer* (2014), https://www.internetretailer.com/commentary/2014/12/04/how-many-online-stores-are-there-world

16   "Internet Usage Statistics: The Internet Big Picture," *Internet World Stats* (2015), http://www.internetworldstats.com/stats.htm

17   Daniel Burrus, "The Internet of Things Is Far Bigger Than Anyone Realizes," *Wired* (2014), https://www.wired.com/insights/2014/11/the-internet-of-things-bigger/

18   "Computer Chips inside the Car," *Chips Etc.* (2016), http://www.chipsetc.com/computer-chips-inside-the-car.html

19   Kevin Kelly, *The Inevitable : Understanding the 12 Technological Forces That Will Shape Our Future* (2016).

20   "Automation," *Oxford English Dictionary*, http://www.oed.com/view/Entry/13468?redirectedFrom=automation#eid

21 Alex Solon, "The Smart Home Gets Smarter," *PC Magazine* (2015), https://www.zinio.com/www/browse/dashboard.jsp?_requestid=130702#/

22 "Device," in *Merriam-Webster.com* (Merriam-Webster, 2017).

23 John R. Patrick, *Health Attitude: Unraveling and Solving the Complexities of Healthcare* (Palm Coast, FL: Attitude LLC, 2015).

24 "Action," in *Merriam-Webster.com* (Merriam-Webster, 2017).

25 Edward C. Papenfuse, "Reading with and About Abraham Lincoln," *Maryland. gov State Archives* (2002), http://msa.maryland.gov/msa/educ/speeches/html/lincoln021502.html

26 T. J. Keefe, "The Nature of Light," *Internet Archive* (2007), https://web.archive.org/web/20120423123823/http://www.ccri.edu/physics/keefe/light.htm

27 Ibid.

28 Joe Romm, "5 Charts That Illustrate the Remarkable LED Lighting Revolution," *ThinkProgress* (2016), https://thinkprogress.org/5-charts-that-illustrate-the-remarkable-led-lighting-revolution-83ecb6c1f472#.bhqfabjks

29 Ibid.

30 Grant Clauser, "Diy or Professional Lighting Control: 5 Tips to Help You Choose," *Electronic House* (2015), https://www.electronichouse.com/home-lighting/diy-or-professional-light-control-5-tips-to-help-you-choose/

31 Ibid.

32 Lisa Montgomery, "19 Awesome Lighting Scenes Your Home Lighting Automation System Can Do," ibid., https://www.electronichouse.com/home-lighting/19-awesome-lighting-scenes-for-your-home/

33 Ibid.

34 "Domino's Fun Facts," *Dominos* (2017), https://biz.dominos.com

35 Ibid.

36 Ibid.

37 "Domino's," *IFTTT* (2017), https://ifttt.com/dominos

38 Ibid.

39 George Rachiotis, "The Importance of Music in Our Daily Lives," (2014), http://www.truthinsideofyou.org/importance-of-music-daily-lives/

40 Ibid.

41 "IFPI Publishes Digital Music Report 2015," *IFPI* (2015), http://www.ifpi.org/news/Global-digital-music-revenues-match-physical-format-sales-for-first-time

42 "MP3 Player," *Wikipedia* (2017), https://en.wikipedia.org/wiki/MP3_player

43 Ibid.

44 "History of iTunes," *Wikipedia* (2017), https://en.wikipedia.org/wiki/History_of_iTunes

45 Ibid.

46 "iPod," *Wikipedia* (2017), https://en.wikipedia.org/wiki/IPod

47 Ibid.

48 "History of Podcasting," *Wikipedia* (2017), https://en.wikipedia.org/wiki/History_of_podcasting

49   "Postwar American Television," *Early Television Museum*, http://www.earlytelevision.org/american_postwar.html

50   Clyde Geronimi, Wilfred Jackson, and Hamilton Luske, "Lady and the Tramp," (1955).

51   Christopher Mele, "Buying a New TV? Here's How to Cut through the Jargon," *The New York Times* (2017), https://www.nytimes.com/2017/03/16/business/buying-a-new-tv-guide.html?ribbon-ad-idx=3&rref=technology&module=Ribbon&version=origin&region=Header&action=click&contentCollection=Technology&pgtype=article

52   "About Niantic, Inc.," *Niantic, Inc.* (2017), https://www.indeed.com/cmp/Niantic,-Inc.

53   Jon Russell, "Report: Pokémon Go Has Now Crossed $1 Billion in Revenue," *TC* (2017), https://techcrunch.com/2017/02/01/report-pokemon-go-has-now-crossed-1-billion-in-revenue/

54   Nick Statt, "Apple Shows Off Breathtaking New Augmented Reality Demos on iPhone 8," *The Verge* (2017), https://www.theverge.com/2017/9/12/16272904/apple-arkit-demo-iphone-augmented-reality-iphone-8

55   David DeMille, "Will Your House Be Broken into This Year?," *ASecureLife* (2017), http://www.asecurelife.com/burglary-statistics/

56   Ibid.

57   "Should I Use Glass Break Detectors or Motion Sensors?," *AlarmGrid* (2017), https://www.alarmgrid.com/faq/should-i-use-glass-break-detectors-or-motion-sensors-to-protect-my-home

58   Richard Davis, "Advantages and Disadvantages of Using Security Cameras," *A1 Security Cameras* (2015), https://www.a1securitycameras.com/blog/advantages-disadvantages-using-security-cameras/

59 "Saving a Life from a Potential Catastrophe," *Life Alert* (2017), http://www.lifealert.com

60 "What You Need to Know About Text-to-911," *Federal Communications Commission* (2016), https://www.fcc.gov/consumers/guides/what-you-need-know-about-text-911

61 Kimberly Hutcherson, "Carbon Monoxide at Hotel Pool Suspected in Michigan Death," *CNN* (2017), http://www.cnn.com/2017/04/01/us/michigan-carbon-monoxide-poisoning/

62 Ibid.

63 "Carbon Monoxide (CO) Detector Buyer's Guide," *SafeWise* (2017), http://www.safewise.com/resources/carbon-monoxide-detectors-guide

64 "CO and CO2 – What's the Difference?," *CO2Meter.com* (2009), https://www.co2meter.com/blogs/news/1209952-co-and-co2-whats-the-difference

65 Ibid.

66 Ibid.

67 Ibid.

68 "Carbon Monoxide at Hotel Pool Suspected in Michigan Death".

69 "What Are the Carbon Monoxide Levels That Will Sound the Alarm?," *Kidde Fire Safety* (2011), http://www.kidde.com/home-safety/en/us/support/help-center/browse-articles/articles/what_are_the_carbon_monoxide_levels_that_will_sound_the_alarm_.aspx

70 "American Housing Survey (AHS)," *U.S. Census Bureau, American Housing Survey* (2015), https://www.census.gov/programs-surveys/ahs/

71   "Kill a Watt® EZ," *P3 International* (2017), http://www.p3international.com/products/p4460.html

72   "Reduce Your Individual Carbon Footprint," *Carbonfund.org* (2016), https://carbonfund.org/individuals/

73   "Aeotec Products," *Aeotec* (2017), https://aeotec.com/

74   "We Make Things Smarter," *sense* (2017), https://sense.com/about.html

75   "Level Indicator Fuel Oil for Fuel Oil Tank, Oil Tank (Wireless) with Separate Radio Display Ecometer," *Proteus* (2016), https://proteus-meter.com/produkte/proteus-ecometer/

76   Philip E. Ross, "Tesla's Model S Will Offer 360-Degree Sonar," *IEEE Spectrum* (2014), http://spectrum.ieee.org/cars-that-think/transportation/self-driving/teslas-model-s-will-offer-360degree-sonar

77   "American Housing Survey (AHS)".

78   Poul-Henning Kamp, "Monitoring Natural Gas Usage," (2001), http://phk.freebsd.dk/Gasdims/

79   Poul-Henning Kamp, Email exchange with author, May 7, 2017.

80   Ibid.

81   "Flow Meter," *FortrezZ* (2017), http://www.fortrezz.com/flow-meter

82   Walt Hickey, "Where People Go to Check the Weather," *FiveThirtyEight* (2015), https://fivethirtyeight.com/datalab/weather-forecast-news-app-habits/

83   Ibid.

84   Ibid.

85   "Experience Connected Weather," *Weather Underground* (2017), https://www.wunderground.com

86   Ibid.

87   "First Complete Z-Wave Weather Station," *Qubino* (2017), http://qubino.com/products/weather-station/

88   "About Ecovacs," *Ecovacs Robotics* (2017), https://www.ecovacs.com/global/about-ecovacs/

89   Chris Savarese and Brian Hart, "The Caesar Cipher," *Cryptography* (2010), http://www.cs.trincoll.edu/~crypto/historical/caesar.html

90   Michael Cavna, "'Nobody Knows You're a Dog': As Iconic Internet Cartoon Turns 20, Creator Peter Steiner Knows the Idea Is as Relevant as Ever," *The Washington Post* (2013), https://www.washingtonpost.com/blogs/comic-riffs/post/nobody-knows-youre-a-dog-as-iconic-internet-cartoon-turns-20-creator-peter-steiner-knows-the-joke-rings-as-relevant-as-ever/2013/07/31/73372600-f98d-11e2-8e84-c56731a202fb_blog.html

91   "Most Common Passwords List," *Random Password generator* (2016), http://www.passwordrandom.com/most-popular-passwords

92   "About Touch ID Security on iPhone and iPad," *Apple* (2016), https://support.apple.com/en-us/HT204587

93   Ibid.

94   Brian X. Chen, "The Smartphone's Future: It's All About the Camera," *The New York Times* (2017), https://www.nytimes.com/2017/08/30/technology/personaltech/future-smartphone-camera-augmented-reality.html?em_pos=large&emc=edit_ct_20170830&nl=technology&nlid=591654&ref=headline&te=1

95   Apple, "About Face ID Advanced Technology," *apple.com* (2017), https://support.apple.com/en-us/HT208108

96   Jake Kouns, "2015 Reported Data Breaches Surpasses All Previous Years," (2016), http://blog.datalossdb.org

97   Lily Hay Newman, "Equifax Officially Has No Excuse," *Wired* (2017), https://www.wired.com/story/equifax-breach-no-excuse/

98   "90% of Data Breaches Are Avoidable," *Cyber Security Intelligence* (2016), https://www.cybersecurityintelligence.com/blog/90-of-data-breaches-are-avoidable-1003.html

99   "Trojan Horse," *Encyclopedia Britannica* (2016), http://www.britannica.com/topic/Trojan-horse

100 "Move over Credit Cards — Stolen Medical Records Are Selling for Record Prices on the Dark Web," *Clearwater Compliance* (2016), https://clearwatercompliance.com/blog/move-credit-cards-stolen-medical-records-selling-record-prices-dark-web/

101 "Samsung SmartThings Hub," *SmartThings* (2017), https://www.smartthings.com/products/samsung-smartthings-hub

102 "Indigo Domotics," *indigo domotics* (2017), http://www.indigodomo.com/

103 John R. Patrick, *Net Attitude : What It Is, How to Get It, and Why You Need It More Than Ever* (Palm Coast, FL: Attitude LLC, 2016).

104 "The Chip That Jack Built," *Texas Instruments* (2016), http://www.ti.com/corp/docs/kilbyctr/jackbuilt.shtml

105 "The First Microprocessor," *AskTheComputerTech.com* (2003), http://www.askthecomputertech.com/first-microprocessor.html

106 Mark Hachman, "Fastest Smartphone Chip Ever,' the A10 Fusion, Powers Apple's New iPhone 7," *Macworld* (2016), http://www.macworld.com/article/3117593/apple-phone/fastest-smartphone-chip-ever-the-a10-fusion-powers-apples-new-iphone-7.html

107 "LED Strip," *Aeotec* (2017), https://aeotec.com/z-wave-led-light-strip

108 "Domino's".

109 Andrew Liptak, "Amazon's Alexa Started Ordering People Dollhouses after Hearing Its Name on TV," *The Verge* (2017), https://www.theverge.com/2017/1/7/14200210/amazon-alexa-tech-news-anchor-order-dollhouse

110 Tim Urban, "The AI Revolution: The Road to Superintelligence," *Wait But Why* (2015), https://waitbutwhy.com/2015/01/artificial-intelligence-revolution-1.html

111 Ibid.

112 Patrick, *Net Attitude: What It Is, How to Get It, and Why Your Company Can't Survive without It.*

113 "Wemo® Switch Smart Plug," *Belkin.com* (2017), http://www.belkin.com/us/F7C027-Belkin/p/P-F7C027;jsessionid=DB246DC0A1F089FC82D3435E89FB3804/?gclid=CjsKDwjw6qnJBRDpoonDwLSeZhIkAIpTR8JNt_8rly5nKWCKXt7WyjPIOEsmNVIkIPTCGQ14jK0GGgJ7NvD_BwE&gclsrc=aw.ds

114 "King of His Digital Castle," *Bloomberg Businessweek* (2007), https://www.bloomberg.com/news/articles/2007-01-21/king-of-his-digital-castle

115 "Home Automation System Market by Protocol and Technology," *MarketsandMarkets* (2017), http://www.marketsandmarkets.com/Market-Reports/home-automation-control-systems-market-469.html?gclid=CjwKEAjwutXIBRDV7-SDvdiNsUoSJACIlTqlw8OMPm05EFZyG_jrZfVZ7y_NoL38qZlN4vd6EiagvhoCBvTw_wcB

116 "Control4 Corporation (Ctrl)," *Yahoo! Finance* (2017), https://finance.yahoo.com/quote/CTRL?p=CTRL

117 "Crestron," *Crestron* (2017), http://crestron.com

118 "Leviton Today," *Leviton.com* (2017), http://www.leviton.com/en/about/about-us/leviton-history/leviton-today

119 "Lutron Services—an Overview," *Lutron.com* (2015), http://www.lutron.com/TechnicalDocumentLibrary/3672404.pdf

120 "About Savant," *Savant.com* (2017), https://www.savant.com/company-info

121 Kimberly Alt, "Frontpoint Security Reviews: Our Top Pick in 2017 Explained," *ASecureLife* (2017), http://www.asecurelife.com/frontpoint-security-reviews/

122 John R. DeLaney, "Frontpoint Home Security System," *PC Magazine* (2017), http://www.pcmag.com/review/351648/frontpoint-home-security-system

123 Andrew Gebhart and Megan Wollerton, "Apple's Home App Makes It Easy to Control Your Home from Your Phone. Finally," *CNET* (2016), https://www.cnet.com/news/exploring-apples-home-for-homekit/

124 "HomeSeer," *HomeSeer* (2017), http://homeseer.com/about-us.html

Index

Y
Yahoo, 155

Z
Z-Wave, 90, 94, 96
Z-Wave protocols, 94
Z-wave vibration sensor, 112
Z-Wave Water Leak Detector, 98

91567484R00117

Made in the USA
Columbia, SC
23 March 2018